SYSTEM MODELING
AND
SIMULATION

SYSTEM MODELING AND SIMULATION

V.P. Singh
Head, Computer Science & Engineering
Institute of Engineering & Technologies
Bhaddal, Ropar (Punjab), India

PUBLISHING FOR ONE WORLD

NEW AGE INTERNATIONAL (P) LIMITED, PUBLISHERS
New Delhi • Bangalore • Chennai • Cochin • Guwahati • Hyderabad
Jalandhar • Kolkata • Lucknow • Mumbai • Ranchi
Visit us at **www.newagepublishers.com**

Published by New Age International (P) Ltd., Publishers
First Edition: 2009

Branches:

- 36, Malikarjuna Temple Street, Opp. ICWA, Basavanagudi, **Bangalore**. ✆ (080) 26677815
- 26, Damodaran Street, T. Nagar, **Chennai**. ✆ (044) 24353401
- Hemsen Complex, Mohd. Shah Road, Paltan Bazar, Near Starline Hotel, **Guwahati**. ✆ (0361) 2543669
- No. 105, 1st Floor, Madhiray Kaveri Tower, 3-2-19, Azam Jahi Road, Nimboliadda, **Hyderabad**. ✆ (040) 24652456
- RDB Chambers (Formerly Lotus Cinema) 106A, Ist Floor, S.N. Banerjee Road, **Kolkata**. ✆ (033) 22275247
- 18, Madan Mohan Malviya Marg, **Lucknow**. ✆ (0522) 2209578
- 142C, Victor House, Ground Floor, N.M. Joshi Marg, Lower Parel, **Mumbai**. ✆ (022) 24927869
- 22, Golden House, Daryaganj, **New Delhi**. ✆ (011) 23262368, 23262370

ISBN : 978-81-224-2386-0

Rs. 495.00

C-08-11-3218

Printed in India at Glorious Printers, Delhi
Typeset In-house

PUBLISHING FOR ONE WORLD

NEW AGE INTERNATIONAL (P) LIMITED, PUBLISHERS
4835/24, Ansari Road, Daryaganj, New Delhi - 110002
Visit us at **www.newagepublishers.com**

PREFACE

Almost half a century has passed since System Analysis emerged as an independent field in Physical Sciences. Number of books and research papers has appeared in the literature and a need is felt to have a systematic one to the study of the subject. The basic techniques of Modeling and Simulation are now being taught in undergraduate engineering courses, and its applications in various engineering subjects require detailed studies. Similarly its application in "Weapon Systems Performance Analysis" is very essential and has also been discussed in the book. Although no claim has been made about the comprehensiveness of this book, yet attempt has been made to make this treatise useful to engineers as well as scientists, especially defence scientists.

The present book is the output of my thirty years of work in the field of Armament and System Analysis and also during my tenure in Institute of Engineering and Technology, Bhaddal, where this subject is being taught at various levels. Most of the chapters in the book are based on the papers published by the author in various technical journals. In order to make the analysis easier to understand, basic mathematical techniques have also been discussed. It is not possible to do full justice to the subject in one small book. But I have tried to condense as much material in this book as possible. I will not say that this treatise is exhaustive, yet it may give quite insight into the subject.

In the end I will like to acknowledge my friend and colleague Sh. Yuvraj Singh, Scientist E., Aeronautical Development Establishment, Bangalore, who was part of my team—Center for Aeronautical System Studies and Analysis, Bangalore and had been quite helpful in preparing this monograph.

V.P. Singh

CONTENTS

WHAT IS A SYSTEM

System modeling and computer simulation, recently has become one of the premier subject in the industry as well as Defence. It helps an engineer or a scientist to study a system with the help of mathematical models and computers. In other words, one can say, system simulation is nothing but an experiment with the help of computers, without performing actual experiment. It saves lot of money which is required, if we actually perform experiments with the real system.

Present book, "System Modeling and Simulation" has been written by keeping engineering students as well as scientists especially defence scientists in mind. Earlier this manuscript was prepared only for the use of defence scientists which has now been extended to other engineering applications. Modeling of a weapon system is just an application of basic techniques of modeling and simulation, which are being discussed in chapter two and four. After my superannuation from Defence Research & Development Organisation, briefly called DRDO in 2000, when I joined the Punjab Technical University, and taught this subject to B. Tech and M. Tech students, the manuscript was rewritten, so that it should be useful to engineering students too. Although many of the examples have been taken from target damage, yet care has been taken to include other examples, from marketing, and mechanical engineering and other related subjects. My intentions are that this book should prove itself as a complete manual for system simulation and modeling. That is the reason, that basic subjects required for modeling and simulation, such as probability theory, numerical methods and C++ have also been included and discussed wherever required. Wherever possible, computer programmes have been given with the output.

First question the user of this book can raise is, after all what is a *system*[1]? We give here a popular definition of what a system is? A broader definition of a system is, "*Any object which has some action to perform and is dependent on number of objects called entities, is a system*". For example a class room, a college, or a university is a system. University consists of number of colleges (which are *entities* of the system called university) and a college has class rooms, students, laboratories and lot many other objects, as *entities*. Each entity has its own attributes or properties. For example attribute of a student is to study and work hard. Each college in itself can be treated as a complete system. If we combine few of these objects, joined in some regular interactions or

1. *Technical terms in italics are indexed at the end of the book.*

inter-dependence, then this becomes a *large system*. Thus we can say university is a large system whereas college is a system. This means, each system can be divided into blocks, where each block is in itself a complete and independently working system (Fig. 0.1). When these blocks are combined, depending on some interdependence, they become entities of a larger system. An aircraft for example, is another example of a system. It consists of a cockpit, pilot, airframes, control system, fuel tank, engines *etc.* Each of these blocks can in itself be treated as a system and aircraft consisting of these interdependent blocks is a large system. Table 0.1 gives components of a system (say college) for illustrations purpose.

Does a system also has some attributes? I will say yes. Systems broadly can be divided into two types, *static system* and *dynamic system*. If a system does not change with time, it is called a *Static System* and if changes with time, it is called a *Dynamic System*. Study of such a system is called *System Analysis*. How this word originated is of interest to know.

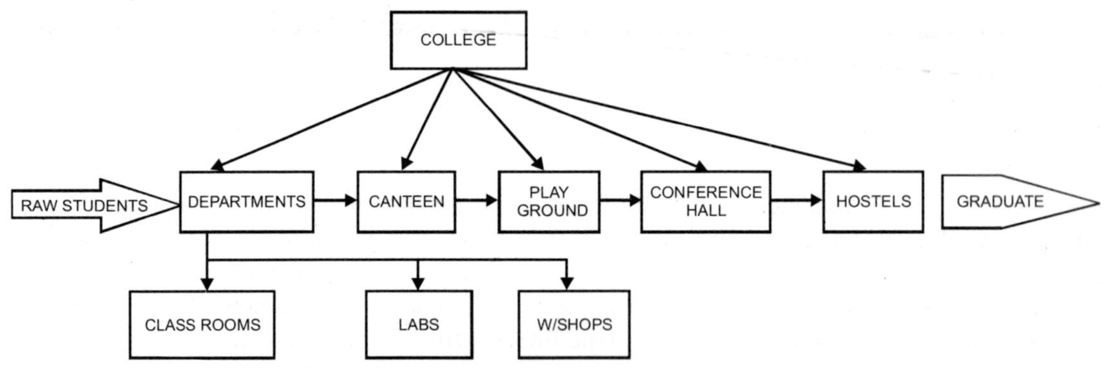

Fig. 0.1: *College as a system.*

While looking at these systems, we see that there are certain distinct objects, each of which possesses some properties of interest. There are also certain interactions occurring in the system that cause changes in the system. A term *entity* will be used to denote an object of interest in a system and the term *attributes* denotes its properties. A function to be performed by the entity is called its *activity*. For example, if system is a class in a school, then students are entities, books are their attributes and to study is their *activity*. In case of the *autopilot* aircraft discussed below, entities of the system are gyroscope, airframe and control surfaces. Attributes respectively are gyroscope setting, speed and control surface angles. Activity of the aircraft is to fly. Banks et al., (2002) has defined state variables and events as components of a system. The state is defined as collection of variables necessary to describe the system at any time, relative to the objectives of the study. An *event* is defined as an instantaneous occurrences that may change the state of the system. But it is felt, entities are nothing but state variables and activity and event are similar. Thus there is no need of further bifurcation.

Sometimes the system is effected by the environment. Such a system is called *exogenous*. If it is not effected by the environment, it is called *endogenous*. For example, the economic model of a country is effected by the world economic conditions, and is exogenous model. Aircraft flight is exogenous, as flight profile is effected by the weather conditions, but static model of the aircraft is endogenous. A class room in the absence of students, is *endogenous*. As mentioned earlier study of a system is called *System Analysis*. How this word "System Analysis" has cropped up? There is an interesting history behind this.

After second world war (1939–1945), it was decided that there should be a systematic study of weapon systems. In such studies, topics like weapon delivery (to drop a weapon on enemy) and weapon assignment (which weapon should be dropped?) should also be included. This was the time when a new field of science emerged and the name given to it was "*Operations Research*". Perhaps this name was after a British Defence Project entitled "Army Operations Research" [19]*. Operations Research at that time was a new subject. Various definitions of "Operations Research" have been given. According to *Morse and Kimball* [1954], it is defined as scientific method of providing executive departments with a quantitative basis for decisions regarding the operations under one's control. According to Rhaman and Zaheer (1952) there are three levels of research : *basic research, applied research and operations research*. Before the advent of computers, there was not much effect of this subject on the weapon performance studies as well as other engineering studies. The reason being that enormous calculations were required for these studies, which if not impossible, were quite difficult. With the advent of computers, "Computerised Operations Research" has become an important subject of science. I have assumed that readers of this book are conversant with numerical methods as well as some computer language.

Table 0.1: Systems and their components

System	*Entities*	*Attributes*	*Activities*
Banking	Customers	Maintaining accounts	Making deposits
Production unit	Machines, workers	Speed, capacity, break-down	Welding, manufacturing
College	Teachers, students	Education	Teaching, games
Petrol pump	Attendants	To supply petrol	Arrival and departure of vehicles

History of Operations Research has beautifully been narrated by Treften (1954). With the time, newer and newer techniques were evolved and name of Operations Research also slowly changed to "System Analysis" [17]. Few authors have a different view that "Operations Research" and "System Analysis" are quite different fields. System Analysis started much later in 1958–1962 during the Kennedy administration in US (Treften, FB 1954). For example, to understand any system, a scientist has to understand the physics of the system and study all the related sub-systems of that system. In other words, to study the performance evaluation of any system, one has to make use of all the scientific techniques, along with those of operations research. For a system analyst, it is necessary to have sufficient knowledge of the every aspect of the system to be analysed. In the analysis of performance evaluation of a typical weapon as well as any other machine, apart from full knowledge of the working of the system, basic mathematics, probability theory as well as computer simulation techniques are used. Not only these, but basic scientific techniques, are also needed to study a system.

In the present book our main emphasis will be on the study of various techniques required for system analysis. These techniques will be applied to various case studies with special reference to weapon system analysis and engineering. Modeling and simulation is an essential part of system analysis. But to make this book comprehensive, wherever possible, other examples will also be given. Present book is the collection of various lectures, which author has delivered in various courses being conducted from time to time for service officers and defence scientists as well as B. Tech

* *Number given in the square bracket* [1], *are the numbers of references at the end of this book. References however are in general given as author name (year of publication).*

and M. Tech students of engineering. Various models from other branches of engineering have also been added while author was teaching at Institute of Engineering and Technology, Bhaddal, Ropar in Punjab. Attempt has been made to give the basic concepts as far as possible, to make the treatise easy to understand.

It is assumed that student, reading this book is conversant with basic techniques of numerical computations and programming in C++ or C language. But still wherever possible, introductory sections have been included in the book to make it comprehensive.

As given earlier, *Simulation* is actually nothing but conducting trials on computer. By simulation one can, not only predict the system behaviour but also can suggest improvements in it. Taking any weapons as an example of a system, lethal capabilities of the weapon which are studied by evaluating its terminal effects i.e., the damage caused by a typical weapon on a relevant target when it is dropped on it, is also one of the major factor. There are two ways of assessing the lethal capabilities of the weapons. One is to actually drop the weapon on the target and see its end effects and the other is to make a model using a computer and conduct a computer trial, which simulates the end effects of the weapon on the target. *Conducting trials on the computers is called computer simulation*. Former way of conducting trial (actual trials) is generally a very costly affair. In this technique, cost of the weapon as well as target, are involved. But the second option is more cost effective.

As an example of a conceptually simple system (Geofrey Gordon, 2004), consider an aircraft flying under the control of an autopilot (Fig. 0.2). A gyroscope in the autopilot detects the difference between the actual heading of aircraft and the desired heading. It sends a signal to move the control surfaces. In response to control surfaces movement, the aircraft steers towards the desired heading.

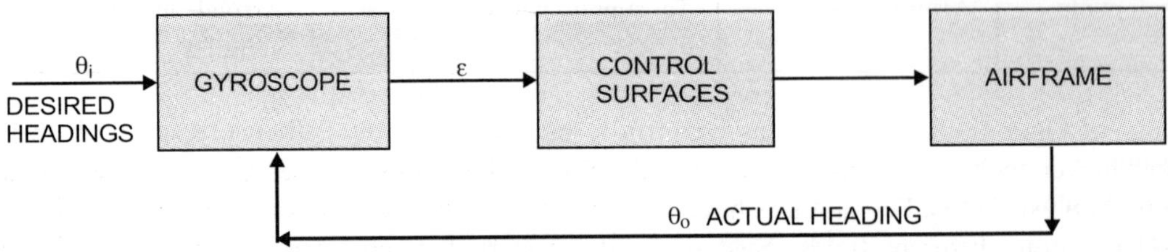

Fig. 0.2: *Model of an autopilot aircraft.*

A factory consisting of various units such as procurement department, fabrication and sale department is also a system. Each department of the factory, on which it depends, is an independent system and can be modelled independently. We have used the word "modelled" here. What is *modeling*, will be discussed in chapter one of this book.

Scientific techniques used in system studies can also broadly be divided into two types:

1. Deterministic studies and
2. Probabilistic studies

Deterministic studies are the techniques, where results are known exactly. One can represent system in the form of mathematical equations and these equations may be solved by analytic methods or by numerical computations. Numerical computations is one of the important tools in system analysis and comes to rescue the system analyst, where analytical solutions are not feasible. In case of study of damage to targets, knowledge of shock waves, fragments penetration, hollow charge and other allied studies is also required. Some of the topics have been introduced in this book for the purpose of broader horizon of the student's knowledge.

Apart from the two types of studies given above, system can be defined as

(*i*) Continuous and

(*ii*) Discrete.

Fluid flow in a pipe, motion of an aircraft or trajectory of a projectile, are examples of continuous systems. To understand continuity, students are advised to refer some basic book on continuity. Examples of discrete systems are, a factory where products are produced and marketed in lots. Motion of an aircraft is continuous but if there is a sudden change in aircraft's level to weather conditions, is a discrete system. Another example of discrete system is firing of a gun on an enemy target.

It is important to conduct experiments to confirm theoretically developed mathematical models. Not only the experiments are required for the validation of the theoretical models but various parameters required for the development of theoretical models are also generated by experimental techniques. For example to study the performance of an aircraft, various parameters like drag, lift and moment coefficients are needed, which can only be determined experimentally in the wind tunnel. Thus theory and experiment both are complementary to each other and are required for correct modeling of a system. In case of marketing and biological models in place of experiments, observations and trends of the system over a time period, are required to be known for modeling the system. This issue will be discussed in chapter eight.

The aim of this book is to discuss the methods of system analysis *vis a vis* an application to engineering and defence oriented problems. It is appropriate here to make reference to the books, by Billy E. Gillett (1979), Pratap K. Mohapatra et al., (1994) and Shannon (1974), which have been consulted by the author.

Layout of the book is arranged in such a way that it is easy for the student to concentrate on examples of his own field and interest. Our attempt is that this book should be useful to students of engineering as well as scientists working in different fields. Layout of the book is as follows:

In **chapter one**, basic concepts of modeling and simulation are given. Number of examples are given to illustrate the concept of modeling. Also different types of models are studied in details.

In **chapter two**, basic *probability theory*, required for this book has been discussed. Probability distribution functions, as used in modeling of various problems have been included in this chapter. Wherever needed, proof for various derivations are also included. Application of probability theory to simple cases is demonstrated as examples. Although it is not possible to include whole probability theory in this book, attempts have been made to make this treatise comprehensive.

Chapter three gives a simple modeling of aircraft survivability, by overlapping of areas, using probability concepts developed in chapter two. It is understood that projection of various parts on a plane are known. This chapter will be of use to scientists who want to learn aircraft modeling tools. However engineering students, if uncomfortable, can skip this chapter.

Simulation by *Monte Carlo technique* using random numbers, is one of the most versatile technique for discrete system studies and has been dealt in **chapter four**. The problems which cannot be handled or, are quite difficult to handle by theoretical methods, can easily be solved by this technique. Various methods for the generation of uniform random numbers have been discussed. Properties of uniform random numbers and various tests for testing the uniformity of random numbers are also given. Normal random numbers are of importance in various problems in nature, including weapon systems. Methods for generating normal random numbers have been discussed in details. Study of different types of problems by computer simulation technique have been discussed in this chapter. Here simulation technique, and generation of different types of random numbers is studied in details. Computer programs in C++ for different techniques of random number generation are also given.

Chapter five deals with the simulation and modeling of *Continuous Dynamic Systems*. Problem becomes slightly complex, when target is mobile. Problem of mobile targets and surface to air weapons, will be studied in chapter six. A study of vulnerability of aerial targets such as aircraft will be discussed. Simulation and modeling can not be learnt without working on practical problems. Therefore number of examples have been worked out in this chapter.

A case study of *aircraft survivability* has been given in **chapter six**. Aircraft model discussed in chapter three was purely on probability concepts. But present model is a simulation model, where all the techniques developed in first five chapters has been utilised. What is the probability that an aircraft will survive when subjected to ground fire in an enemy area, has been studied in this model. This model involves mathematical, and computer modeling. Concept of continuous and stochastic modeling has also been used wherever required.

Simulation of manufacturing process and material handling is an important subject of Mechanical Engineering. *Queuing theory* has direct application in maintenance and production problems. Queuing theory has been discussed in **chapter seven**. Various applications on the field of manufacturing process and material handling have been included in this chapter.

There are various phenomena in nature which become unstable, if not controlled in time. This is possible only if their dynamics is studied and timely measures are taken. Population problem is one of the examples. Nuclear reaction is another example. This type of study is called *Industrial dynamics*. **Eighth chapter** deals with System Dynamics of such phenomena. Various cases of growth and decay models with examples have been discussed in this chapter.

One of the pressing problems in the manufacturing and sale of goods is the control of inventory holding. Many companies fail each year due to the lack of adequate control of inventory. **Chapter nine** is dedicated to inventory control problems. Attempt has been made to model various inventory control conditions mathematically.

Costing of a system is also one of the major job in system analysis. How much cost is involved in design and manufacturing of a system is discussed in **tenth chapter**. Cost effectiveness study is very important whether procuring or developing any equipment. In this chapter basic concepts of costing *vis* a *vis* an application to aircraft industry has been discussed.

EXERCISE

1. Name several entities, attributes, activities for the following systems.
    - A barber shop
    - A cafeteria
    - A grocery shop
    - A fast food restaurant
    - A petrol pump

2. Classify following systems in static and dynamic systems.
   - An underwater tank
   - A submarine
   - Flight of an aircraft
   - A college
   - Population of a country
3. Classify following events into continuous and discrete.
   - Firing of a gun on enemy
   - Dropping of bombs on a target
   - Tossing of a coin
   - Flow of water in a tap
   - Light from an electric bulb.

MODELING AND SIMULATION

Term Modeling and Simulation is not very old and has become popular in recent years. But the field of modeling and simulation is not new. Scientific workers have always been building models for different physical scenarios, mathematically as well as in laboratories. Here question arises, what is the basic difference between modeling and simulation? In fact, there is very thin border line between the both. In fact, physical models can not be called simulation, but mathematical models can be called simulation. Similarly computer simulation can not be given name of modeling but mathematical simulation can be called modeling. Model of a system is the replica of the system, physical or mathematical, which has all the properties and functions of the system, whereas simulation is the process which simulates in the laboratory or on the computer, the actual scenario as close to the system as possible. In fact, a modeling is the general name whereas simulation is specific name given to the computer modeling. Models can be put under three categories, *physical models*, *mathematical models and computer models*. All of these types are further defined as static and dynamic models. Physical model is a scaled down model of actual system, which has all the properties of the system, or at least it is as close to the actual system as possible. Now-a-days small models of cars, helicopters and aircraft are available in the market. These toys resemble actual cars and aircraft. They are static physical models of dynamic systems (cars, helicopters and aircraft are dynamic systems). In wind tunnel, scaled down models of an aircraft are used to study the effect of various aerodynamic forces on it. Similarly before the construction of big buildings, scaled down models of the buildings are made. Well known laws of similitude are used to make the laboratory models. All the dimensions are reduced with respect to some critical lengths (Sedov LI, 1982). Figure 1.1 gives different types of models. These types will be studied in details in the coming chapters.

While building a model certain basic principles are to be followed. While making a model one should keep in mind five basic steps.

- Block building
- Relevance
- Accuracy
- Aggregation
- Validation

A model of a system can be divided into number of blocks, which in itself are complete systems. But these blocks should have some relevance to main system. For example, let us take an example of a school. Class rooms are blocks of the school. Aim of the school is to impart education to students and class rooms are required for the coaching. Thus relevance of class rooms (blocks) with school is coaching. Interdependency is the one of the important factor of different blocks. Each block should be accurate and tested independently. Then these blocks are to be integrated together. Last is the validation i.e., this model is to be tested for its performance. For validation following methods can be used.

- If the model is mathematical model, then some trivial results can be run for verifications.
- If experimental results are available, model can be checked with these experimental results.

In the following sections, we will discuss in details the various types of models as shown in Fig. 1.1.

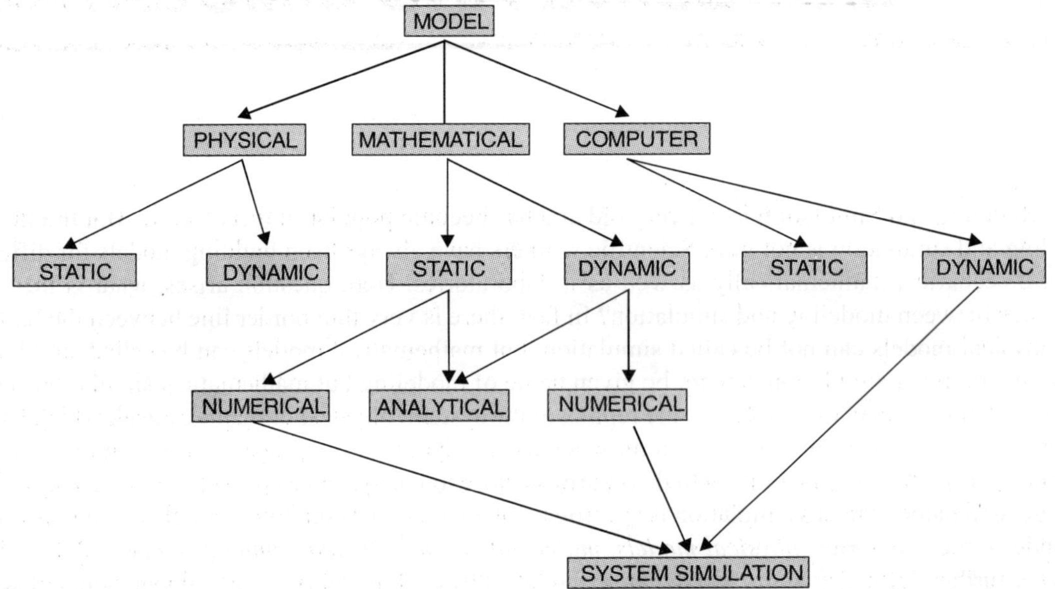

Fig. 1.1: *Different types of models.*

1.1 Physical Models

Physical models are of two types, static and dynamic. Static physical model is a scaled down model of a system which does not change with time. An architect before constructing a building, makes a scaled down model of the building, which reflects all it rooms, outer design and other important features. This is an example of static physical model. Similarly for conducting trials in water, we make small water tanks, which are replica of sea, and fire small scaled down shells in them. This tank can be treated as a static physical model of ocean.

Dynamic physical models are ones which change with time or which are function of time. In wind tunnel, small aircraft models (static models) are kept and air is blown over them with different velocities and pressure profiles are measured with the help of transducers embedded in the model. Here wind velocity changes with time and is an example of dynamic physical model. A model of a hanging wheel of vehicle is another case of dynamic physical model discussed further.

Let us take an example of hanging wheel of a stationary truck and analyze its motion under various forces. Consider a wheel of mass M, suspended in vertical direction, a force $F(t)$, which varies with time, is acting on it. Mass is connected with a spring of stiffness K, and a piston with damping factor D. When force $F(t)$, is applied, mass M oscillates under the action of these three forces. This model can be used to study the oscillations in a motor wheel. Figure 1.2 shows such a system. This is a discrete physical static model. Discrete in a sense, that one can give discrete values F and observe the oscillations of wheel with some measuring equipment. When force is applied on it, which is a function of time, this discrete physical static model becomes dynamic model. Parameters K and D can also be adjusted in order to get controlled oscillations of the wheel. This type of system is called spring-mass system. Load on the beams of a building can be studied by the combination of spring-mass system. Mathematical model of this system will be studied in coming chapters.

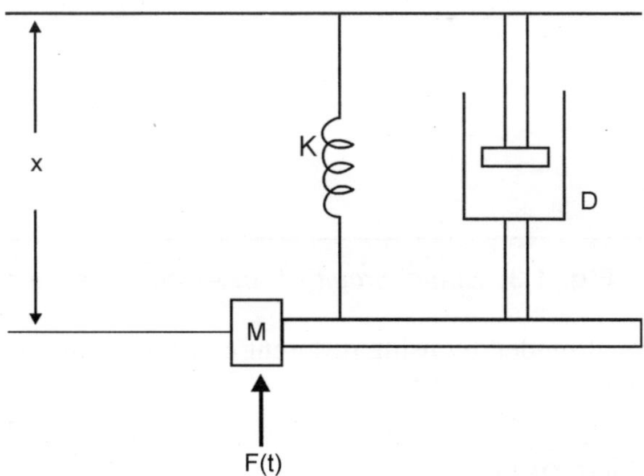

Fig. 1.2: *Suspended weight attached with spring and piston.*

Let us consider another static physical model which represents an electric circuit with an inductance L, a resistance R, and a capacitance C, connected with a voltage source which varies with time, denoted by the function $E(t)$. This model is meant for the study of rate of flow of current as $E(t)$ varies with time. There is some similarity between this model and model of hanging wheel. It will be shown below that mathematical model for both is similar. These physical models can easily be translated into a mathematical model. Let us construct mathematical model of system describing hanging wheel. Using Newton's second law of motions, system for wheel model can be expressed in the mathematical form as

$$M\frac{d^2x}{dt^2} + D\frac{dx}{dt} + Kx = KF(t) \qquad \text{...(1.1)}$$

where x = the distance moved,

M = mass of the wheel,

K = stiffness of the spring,

D = damping force of the shock absorber.

This system is a function of time and is a dynamic system. Equation (1.1) cannot be solved analytically and computational techniques can be used for solving the equations. Once we interpret the results of this equation, this becomes a dynamic mathematical model of the system. Physical model can also be used to study the oscillations by applying force F, and measuring the displacement x of the model. Values of K and D can be changed for required motion with minimum oscillations. This equation will be discussed in the section on Mathematical models in section 1.2. Equation of electrical circuit given above can be written as

$$L\ddot{q} + R\dot{q} + \frac{q}{C} = \frac{E(t)}{C}$$

Here the $\dot{q}$ is the electric current. In this equation if we say $q = x$, $L = M$, $R = D$, $E = F$ and $K = 1/C$, then one can easily see that this equation is similar to (1.1).

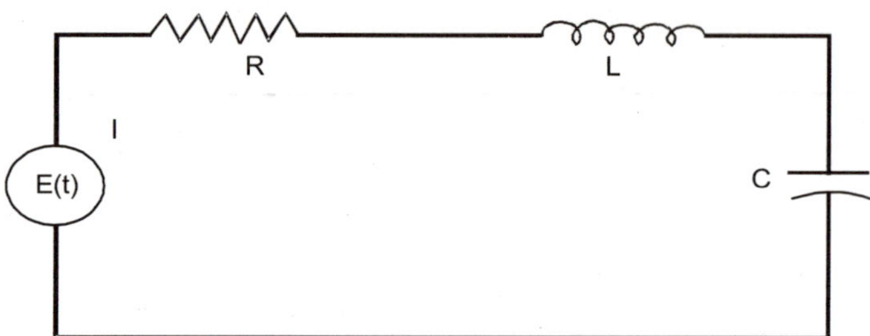

Fig. 1.3: *Static circuit of an electrical system.*

Thus same mathematical model, by using different constants can give solution for hanging wheel as well as electrical circuit.

1.2 Mathematical Models

In section 1.1 we have seen, how a physical model can be converted to mathematical model. Most of the systems can in general be transformed into mathematical equations. These equations are called the mathematical model of that system. Since beginning, scientists have been trying to solve the mysteries of nature by observations and also with the help of Mathematics. Kepler's laws represent a dynamic model of solar system. Equations of fluid flow represent fluid model which is dynamic. A static model gives relationships between the system attributes when the system is in equilibrium. Mathematical model of a system, in equilibrium is called a *Static Mathematical Model.* Model of a stationary hanging wheel equation (1.1) is a dynamic mathematical model, as equations of the model are function of time. This equation can be solved numerically with the help of Runge-Kutta method.

It is not possible to find analytic solution of this equation and one has to adopt the numerical methods. We divide equation (1.1) by M and write in the following form (Geofrey Gordon, 2004)

$$\frac{d^2x}{dt^2} + 2\zeta\omega\frac{dx}{dt} + \omega^2 x = \omega^2 F(t) \qquad ...(1.2)$$

where $2\zeta\omega = D/M$ and $\omega^2 = K/M$. Expressed in this form, solution can be given in terms of the variable ωt. Details of the solution of this equation will be given in chapter five on continuous models. Expressed in this form, solution can be given in terms of the variable ωt, where ω is the frequency of oscillation given by

$$\omega^2 = 2\pi f$$

and f is the number of cycles per second.

We can integrate equation (1.2) using numerical techniques. It will be shown that wheel does not oscillate for $\zeta \geq 1$.

1.2.1 Static Mathematical Models

Mathematical model of equation (1.2) involves time *t*, and is thus dynamic model. If mathematical model does not involve time i.e., system does not change with time, it is called a static mathematical model of the system. In this section, we will discuss few static mathematical models. First of the model is for evaluating the total cost of an aircraft sortie, for accomplishing a mission. Costing is one of the important exercise in System Analysis, and is often required for accomplishing some operations. Second case has been taken from the market. A company wants to optimize the cost of its product by balancing the supply and demand. In a later section, this static model will be made dynamic by involving time factor in it. A student training model, in which marks allotted to students are optimized, is also presented.

1.2.2 Costing of a Combat Aircraft

This section can be understood in a better way after chapter ten. But since this problem falls under static modeling, it is given here. Students of engineering can skip this section if they like. In this model we assume that a sortie of fighter aircraft flies to an enemy territory to destroy a target by dropping the bombs. Now we will construct a mathematical model to evaluate the total cost of attack by aircraft sortie to inflict a stipulated level of damage to a specified target. This cost is the sum of (*a*) Mission cost of surviving aircraft, (*b*) Cost of aborted mission, (*c*) Cost of killed aircraft/repair cost, (*d*) Cost of killed pilots, and (*e*) Cost of bombs. Here surviving aircraft mean, the number of aircraft which return safely after attacking the enemy territory. Aborted mission means, aircraft, which could not go to enemy territory due to malfunctioning of aircraft. Meaning of other terms is clear.

(a) Mission cost of surviving aircraft

In order to evaluate these costs, we have to know, how many aircraft have survived after performing the mission and how many have been killed due to enemy firing. Let *N* be the total number of aircraft sorties which are sent on a mission, then

$$N = N_w / [N_b \cdot p \cdot p_1 \ (1 - p_a) \ (1 - p_b) \ (1 - p_c)] \qquad \text{...(1.3)}$$

where

N_w = number of bombs required to damage the given target up to the specified level,
N_b = number of bombs per one aircraft sortie,
p = survival probability of aircraft in enemy air defence environments,
p_1 = weapon reliability,
p_a = abort probability before reaching the target,
p_b = abort probability on account of not finding the target,
p_c = abort probability due to failure of release mechanism.

It is assumed that the various probabilities in equation (1.3) are known based on earlier war experience.

Due to various reasons, it is quite possible that some of the aircraft are not able to perform mission during operation, and thus abort without performing the mission. To take such aborted aircraft into account, let N_a, N_b and N_c be the number of aircraft aborted due to some malfunctioning, due to missing the target and due to failure of weapon release mechanism respectively. Then N_a, N_b and N_c are given as follows:

$$N_a = N \cdot p_a$$
$$N_b = N \cdot (1 - p_a)\, p_b \qquad ...(1.4)$$
$$N_c = N \cdot (1 - p_a)\,(1 - p_b)\, p_c$$

In equation (1.4), N_b is the product of aircraft not aborted due to malfunctioning i.e., $N(1 - p_a)$ and probability of abort due to missing the target (p_b). Same way N_c is evaluated. Since N_a number of aircraft have not flown for the mission, we assume, that they are not subjected to enemy attrition. Then total number of aircraft going for mission is

$$N_1 = N - N_a = N(1 - p_a) \qquad ...(1.5)$$

N_b and N_c are number of aircraft being aborted after entering into enemy territory and thus are subject to attrition. Therefore total number of aircraft accomplishing the successful mission is

$$N - N_a - N_b - N_c$$

Out of which surviving aircraft are given as

$$N_s = p(N - N_a - N_b - N_c)$$

or by using (1.4) in the above equation,

$$N_s = p \cdot N\,[1 - p_a - (1 - p_a)\,p_b - (1 - p_a)\,(1 - p_b)\,p_c] \qquad ...(1.6)$$

There are another category of surviving aircraft (given by equation (1.4)) which have not been able to accomplish the mission. Amongst these N_a aircraft aborted before entering the enemy territory and are not subject to attrition, where as N_b and N_c category have attrition probability p, as they abort after reaching near the target. Let these aircraft be denoted by N_{as}. Then

$$N_{as} = N_a + (N_b + N_c) \cdot p \qquad ...(1.7)$$

Therefore total number of survived aircraft is

$$N_s + N_{as}$$

Once number of survived aircraft is evaluated, we can fix a cost factor. This is accomplished as follows.

If the acquisition cost of an aircraft, whose total life in H hours is C, then the life cycle cost of the aircraft, is given by

$$C(1 + k)$$

where k is the cost factor for computation of life cycle cost which includes cost of spares and maintenance, operation etc., and strike-off wastage during its life. Value of cost for computation of life cycle cost which includes cost of various aircraft is given in Appendix 10.1. The cost of the surviving aircraft for the assigned mission, and have performed the mission is given by

$$C_1 = (C(1 + k)\, N_s t/H), \qquad ...(1.8)$$

where t is mission sortie time in hours.

(b) Cost of aborted mission (C_2)

The number of surviving aborted aircraft as given in equation (1.7), are N_{as}. Thus the total cost of such aborted aircraft is

$$C_2 = \frac{N_a C(1+k) \cdot t_a}{H} + \frac{p \cdot N_b \cdot C(1+k) \cdot t_b}{H} + \frac{p \cdot N_c \cdot C(1+k) \cdot t_c}{H} \qquad ...(1.9)$$

where t_a, t_b, t_c are the sortie time taken in each of the aborted cases mentioned above. Timings t_a, t_b and t_c are input data and is based on the war estimates and total sortie time t of the aircraft. For example $t_c = t$ as in this case the aircraft had reached the target but could not release the weapon due to failure of firing mechanism. Similarly t_b can also be taken as equal to t as aircraft has flown up to the target but could not locate it.

(c) Cost of killed aircraft

Let N_k be the total number of killed aircraft given by $N_k = (N - N_a)\cdot(1 - p)$ and t_j is the life already spent by N_j-th aircraft, then the total cost of killed aircraft is given by

$$C_3 = \frac{C}{H}\left(1 + \frac{C_{sow}}{C}\right)\sum_{j=1}^{p} N_j (H - t_j) \quad ...(1.10)$$

where $N_k = N_1 + N_2 + \cdots + N_p$ and the terms C_{sow}/C in above equation has been taken due to the reasons that when an aircraft which has spent a life of t_j hours is lost, partial cost of spares, fuel, and maintenance is not lost. Only loss in this case is $C + C_{sow.}$ Thus if we assume that all the aircraft are new i.e., $t_j = 0$, we have,

$$C_3 = N_k C\left(1 + \frac{C_{sow}}{C}\right) \quad ...(1.11)$$

It is to be noted that in this equation only cost due to strike-off wastage is taken out of cost factor k. This is because cost of spares and maintenance is not assumed as gone when an aircraft is killed.

(d) Cost of killed pilots

Let p_2 be pilot survival probability of killed aircraft. Then the cost of the killed pilots is given by

$$C_4 = (1 - p_2)\ \text{int}\ (N_k) \cdot C_p \quad ...(1.12)$$

where C_p is the cost of killed pilot and int (N_k) means *integer part* of $N_{k.}$ Cost of pilot is based on the total expenditure of giving him training and compensation involved in case of his death.

(e) Cost of bombs

If C_b is the cost of a bomb, then the total cost of bombs is given by

$$C_5 = \frac{N_w C_b}{p_1} \quad ...(1.13)$$

where N_w are the total number of bombs required to damage the target and p_1 is the reliability that bomb will function.

Thus the total cost of the attack by aircraft to inflict a stipulated level of damage to a specified target is the sum of all the individual costs defined above and is given by

$$C_{a/c} = C_1 + C_2 + C_3 + C_4 + C_5 \quad ...(1.14)$$

1.2.3 A Static Marketing Model

In order to illustrate further, we take one more example of static mathematical model. Here we give a case of static mathematical model from industry. Generally there should be a balance between the supply and demand of any product in the market. Supply increases if the price is higher. This is because shopkeeper gets more commission on that product and tries to push the product to the customers even if quality is not excellent. Customer generally feels that more cost means better quality. But on the other hand demand decreases with the increase of price. Aim is to find the optimum price with which demand can match the supply. Let us model this situation mathematically. If we denote price by P, supply by S and demand by D, and assuming the price equation to be linear we have

$$D = a - bP$$
$$S = c + dP \qquad ...(1.15)$$
$$S = D$$

In the above equations, a, b, c, d are parameters computed based on previous market data. Equation $S = D$ says supply should be equal to demand so that market price should accordingly be adjusted. Let us take values of $a = 500$, $b = 2000$, $c = -50$ and $d = 1500$. Value of c is taken negative, since supply cannot be possible if price of the item is zero. In this case no doubt equilibrium market price will be

$$P = \frac{a-c}{b+d} = \frac{550}{3500} = 0.1571$$

and $S = 186$

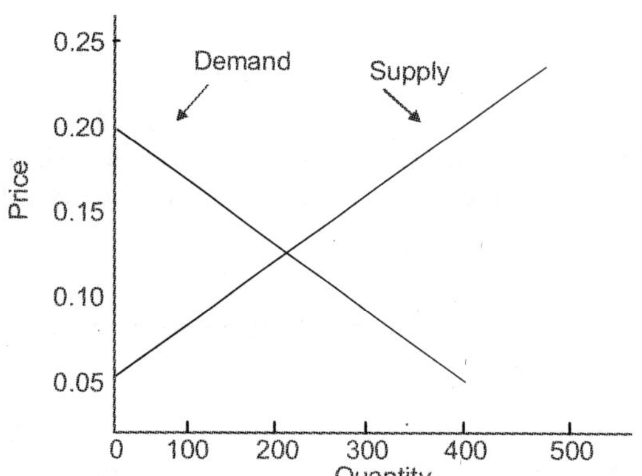

Fig. 1.4: *Market model.*

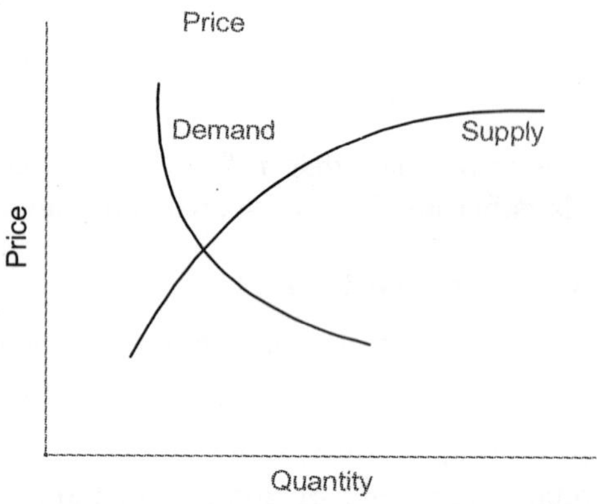

Fig. 1.5: *Non-linear market model.*

In this model we have taken a simplistic linear case but equation (1.15) may be complex. In that case solution may not be so simple. More usually, the demand and supply are depicted by curves with slopes downward and upward respectively (Fig. 1.5). It may not be possible to express the relationships by equations that can be solved easily. Some numerical or graphical methods are used to solve such relations. In addition, it is difficult to get the values of the coefficients of the model. Observations over the extended period of time, however, will establish the slopes (that is values of b and d) in the neighbourhood of the equilibrium points. These values will often fluctuate under the global and local economic conditions.

1.2.4 Student Industrial Training Performance Model

For engineering students, six months training in industry is a part of their course curriculum and is compulsory. It has been observed that training marks allotted to students from industrial Institutions, vary drastically, irrespective of the academic record of the student. Nature of project offered and standard of training institutions pay a very dominating role in these criteria. Due to this sometimes, very good students, who are supposed to top the University exam, can suffer with no fault on their part. A model to optimize the industrial marks which are about 40% of total marks is presented below.

Let M be the marks allotted by industry and $\overline{M}$ the optimized marks.

If m = average result of the college.

m_{ac} = % average marks of student in first to six semester.

m_{ext} = % marks of training given by external expert.

m_{int} = % marks given by internal evaluation.

m_{ind} = % average marks given by industry for last two years.

w_{ac} = weight factor of academic.

w_{ext} = weight factor of external marks.

w_{int} = weight factor of internal marks.

w_{ind} = weight factor of industry.

Then,

$$\overline{M} = M + w_{ac}\,(m_{ac} - m) + w_{ext}\,(m_{ext} - m) + w_{int}\,(m_{int} - m) - w_{ind}\,(m_{ind} - m)$$

minus sign to industry is because it is assumed that good industry does not give high percentage of marks.

Weight factors are as follows:

$w_{ac} = 0.5$

$w_{ext} = 0.25$

$w_{int} = 0.125$

$w_{ind} = 0.125$

$m = 0.8$ (say)

(I) Case Study (**Good Student**)

$M = 0.75$

$m_{ac} = 0.95$

$m_{ext} = 0.9$

$m_{int} = 0.95$

$m_{ind} = 0.6$

Then,

$$\overline{M} = 0.75 + 0.5 \times (0.15) + 0.25 \times (0.10) + 0.125 \times (0.15) - 0.125 \times (0.6 - 0.8)$$

$$\Rightarrow \quad \overline{M} = 0.75 + 0.075 + 0.025 + 0.019 + 0.025$$

$$\therefore \quad \overline{M} = 0.894$$

(II) Case Study (**Poor Student**)

$M = 0.95$

$m_{ac} = 0.50$

$m_{ext} = 0.6$

$m_{int} = 0.5$

$m_{ind} = 0.9$

Then,

$$\overline{M} = 0.95 + 0.5 \times (0.5 - 0.8) + 0.25 \times (0.6 - 0.8) + 0.125 \times (0.5 - 0.8) - 0.125 \times (0.9 - 0.8)$$

$$\Rightarrow \overline{M} = 0.95 + 0.15 - 0.05 - 0.034 - 0.0125$$

$$\therefore \overline{M} = 0.70$$

Model can be used for upgrading or down grading (if required) the industrial marks.

1.3 Computer Models

With the advent of computers, modeling and simulation concepts have totally been changed. Now all types of stochastic as well as continuous mathematical models can be numerically evaluated with the help of numerical methods using computers. Solution of the problem with these techniques is called computer modeling. Here one question arises, what is the difference between mathematically obtained solution of a problem and *simulation*. Literal meaning of simulation is to simulate or copy the behavior of a system or phenomenon under study. Simulation in fact is a computer model, which may involve mathematical computation, computer graphics and even discrete modeling. But Computer oriented models which we are going to discuss here are different from this. One can design a computer model, with the help of graphics as well as mathematics, which simulates the actual scenario of war gaming. Let us assume, we have to model a war game in which missile warheads are to be dropped on an airfield. Each warhead is a cratering type of warhead which makes craters on the runway so that aircraft can not take off. This type of warhead is also called Blast Cum Earth Shock (BCES) type of warhead.

1.3.1 Runway Denial using BCES Type Warhead

First requirement of air force during war time, is to make the enemy's runway unserviceable, so that no aircraft can take off or land on it. It has been observed that during emergency, a modern fighter aircraft like F-16, is capable of taking off even if a minimum strip of 1000m × 15m is available. Thus in order that no where on airstrip, an area of 1000m × 15m is available, bombs which create craters on the runway are dropped by the aircraft or missile. This model was first developed by the author, when a requirement by army was sent to Center for Aeronautical System Studies and Analysis, Bangalore to study the cost-effectiveness of *Prithvi* vs. deep penetrating aircraft. *Prithvi* is a surface to surface missile, developed by Defence Research and Development Organization. Blast Cum Earth Shock warhead (BCES), generally used against runway tracks is capable of inflicting craters to the tracks and making them unserviceable. These types of warheads earlier used to be dropped on runway in the pattern of stick bombing (bombs dropped in a line). In this section a simulation and mathematical model for the denial of an airfield consisting of runway tracks inclined at arbitrary angles, using BCES type of warheads has been discussed.

A Static model of airfield: In this section we construct a static model of an airfield. An airfield consisting of three tracks (a main runway denoted by 'RW' and a taxi track denoted by 'CW' and another auxiliary runway denoted by 'ARW') is generated by computer using graphics (Fig. 1.6). The airfield consists of one runway (3000 × 50m), a parallel taxi track (3000 × 25m) and an auxiliary runway normal to both (2500 × 50m). This model is developed to simulate the denial of runway by dropping bombs on it. The denial criterion of the airfield is that, no where on the track, a strip of dimensions 1000 × 15m, which is sufficient for an aircraft to take off in emergency, is available. To check whether this track (1000 × 15) square meters is available or not, an algorithm has been developed by Singh et al., (1997) devising a methodology for checking the denial criterion.

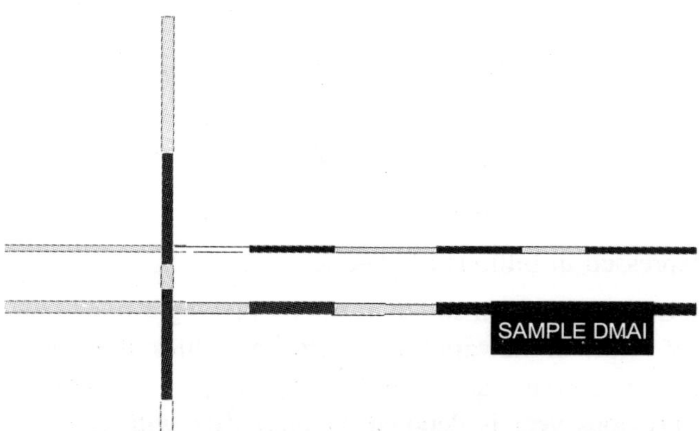

Fig. 1.6: *A typical airfield, showing runways and DMPIs (black sectors).*

To achieve this, we draw certain areas on the track so that if atleast one bomb falls on each of these areas, 1000 meters will not be available anywhere on the three tracks. These areas are called Desired Mean Areas of Impact (DMAI), and have been shown as black areas in Fig. 1.6. Desired Mean Points of Impact (DMPIs) and strips are chosen in such a way that, if each strip has at least one bomb, no where a strip of dimensions 1000m × 15m is available. Number of strips N_s of effective width W_s in a DMAI is given by,

$$N_s = \begin{cases} 1, & \text{if } W_d = W \\ \text{int}\left(\dfrac{2W}{W_d + 2r_b}\right) + 1, & \text{otherwise} \end{cases} \qquad ...(1.16)$$

where W, W_d, r_b are the width, denial width respectively of the RW and lethal radius of the bomblet.

Monte Carlo computer model of airfield denial has been discussed by Singh et al., (1997) and is given in chapter four.

1.3.2 Distributed Lag Models—Dynamic Models

The market model, discussed in section 1.2.3 was straight forward and too simplistic. When model involves number of parameters and hefty data, one has to opt for computer. Models that have the properties of changing only at fixed intervals of time, and of basing current values of the variables on other current values and values that occurred in previous intervals, are called *distributed lag models* [Griliches, Zvi 1967]. These are a type of dynamic models, because time factor is involved in them. They are extensively used in econometric studies where the uniform steps correspond to a time interval, such as a month or a year, over which some economic data are collected. As a rule, these model consists of linear, algebraic equations. They represent a continuous system, but the one in which data is only available at fixed points in time.

As an example, consider the following simple dynamic mathematical model of the national economy. Let,

C be consumption,

I be investment,

T be taxes,

G be government expenditure and Y be national income.

Then

$$\left.\begin{aligned} C &= 20+0.7(Y-T) \\ I &= 2+0.1Y \\ T &= 0+0.2Y \\ Y &= C+I+G \end{aligned}\right\} \qquad \text{...(1.17)}$$

All quantities are expressed in billions of rupees.

This is a static model, but it can be made dynamic by picking a fixed time interval, say one year, and expressing the current values of the variables in terms of values of the previous year. Any variable that appears in the form of its current value and one or more previous year's values is called *lagged variables*. Value of the previous year is denoted by the suffix with-1.

The static model can be made dynamic by lagging all the variables, as follows;

$$\left.\begin{aligned} C &= 20+0.7(Y_{-1}-T_{-1}) \\ I &= 2+0.1Y_{-1} \\ T &= 0.2Y_{-1} \\ Y &= C_{-1}+I_{-1}+G_{-1} \end{aligned}\right\} \qquad \text{...(1.18)}$$

In these equations if values for the previous year (with –1 subscript) is known then values for the current event can be computed. Taking these values as the input, values for the next year can also be computed. In equation (1.18) we have four equations in five unknown variables.

It is however not necessary to lag all the variable like it is done in equation (1.18). Only one of the variable can be lagged and others can be expressed in terms of this variable. We solve equation for Y in equation (1.17) as

$$\begin{aligned} Y &= 20 + 0.7(Y - 0.2Y) + I + G \\ &= 20 + 0.56Y + I + G \end{aligned}$$

or

$$Y = 45.45 + 2.27(I + G)$$

Thus we have,

$$\left.\begin{aligned} I &= 2.0+0.1Y_{-1} \\ Y &= 45.45+2.27(I+G) \\ T &= 0.2Y \\ C &= 20+0.7(Y-T) \end{aligned}\right\} \qquad \text{...(1.19)}$$

In equations (1.19) only lagged parameter is Y. Assuming that government expenditure is known for the current year, we first compute I. Knowing I and G, Y and T for the current year is known, and thus C is computed from the last equation. In this problem, lagged model is quite simple and can be computed with hand calculator. But national economic models are generally not that simple and require long computations with number of parameters.

1.4 Cobweb Models

In section 1.2.3, a simple static model of marketing a product had been discussed. In that model two linear equations for demand D and supply S were considered. Aim was to compute the probable price and demand of a product in the market subject to a condition that supply and demand should be equal. But supply of the product in the market depends on the previous year price, and that can be taken as lagged parameter. Thus equations (1.15) become

$$D = a - bP$$
$$S = c + dP_{-1} \qquad \text{...(1.20)}$$
$$D = S$$

Model 1	*Model 2*
P_0 = 30	P_0 = 5
a = 12	a = 10.0
b = 30	b = 0.9
c = 1.0	c = –2.4
0.9	1.2

In the equations (1.20), values of parameters *a*, *b*, *c*, and *d* can be obtained by fitting a linear curve in the data from the past history of the product. We assign some initial value to the product say P_0 and find *S* from second equation of (1.20). Thus *S* and *D* are known and first equation of (1.20) gives us new value of *P*. Using this value of *P* as initial value, we repeat the calculations and again compute *P* for the next period. If price converges, we say model (1.20) is stable. Let us take two examples and test whether these models converge or not.

Table 1.1: Cobweb model for marketing a product

Model 1		*Model 2*	
i	*P*	*i*	*P*
0	–0.533333	0	7.11111
1	0.382667	1	4.2963
2	0.355187	2	8.04938
3	0.356011	3	3.04527
4	0.355986	4	9.71742
5	0.355987	5	0.821216
6	0.355987	6	12.6828
7	0.355987	7	–3.13265
8	0.355987	8	17.9546
9	0.355987	9	–10.1618
10	0.355987	10	27.3268

We have given a small program to find the value of *P* on the next page.

Program 1.1: Program for Cobweb Model for Marketing a Product

```
# include <fstream.h>
void main (void)
{
  double po, a,b,c,d,s,q,p;
  int i;
  ofstream outfile;
  outfile.open("vps");
  cout << " type values of Po, a, b, c, d\n";
       cin >>po >>a>>b>>c>>d;
    cout <<po<<'\t'<<a<<'\t'<<b<<'\t'<<c<<'\t'<<d<<'\t'<<"\n";
       outfile <<"i"<<'\t' <<"po" <<"\n";
       for (i=0; i<20; i++)
       {
       s  =  c +d*po;
       q  =  s;
       p  =  (a -q)/b;
       po= p;
    outfile <<i<<'\t' <<po <<"\n";
       }
   }
```

results of two models are given in the Table 1.1.

We can see from the table that results in the case of first model converge even in five steps where as in second model they do not converge et al. Data *a*, *b*, *c*, and *d* for model 2 is such that it does not converge. Thus data of second model is not realistic. These parameters can be calculated from the past history of the product by regression method. This model is called cobweb as it can be graphically expressed as shown in Fig. 1.7.

In Fig. 1.7, we have first drawn supply and demand curves. A line parallel to quantity axis shows that for price equal to one unit, supply is 2 units. If we draw a line parallel to price axis so that it meets demand curve at point marked 1. Thus for the same quantity of supply and demand, price immediately shoots up to more than eight units, due to short supply of product. With this high price, supply shoots up to nine units. Again vertical line equating supply with demand reduces the price to three. We repeat the process and ultimately find that curve converges to optimum value.

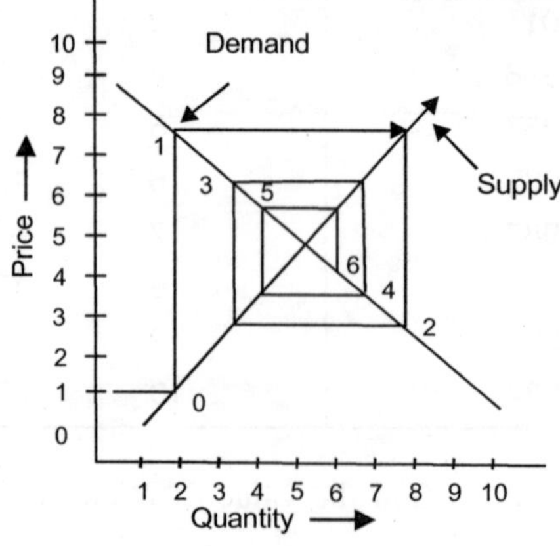

Fig. 1.7: *Cobweb model.*

For simulating the real life systems, no real science can serve the purpose, because knowledge of different discipline is generally needed. That is why sometimes simulation is called an art and not science. In the coming chapters, we will study different techniques, required for simulation. Probability theory is one of the important scientific fields required for stochastic simulation. In next chapter, we will study probability in details.

1.5 SIMULATION

In the section 1.2, we had given an example of a mathematical model of hanging wheel of a vehicle. It will be shown by numerical computations of the equation (1.1) in chapter five, that system does not oscillate when parameter ζ is greater than or equal to one. This we could find by numerically integrating the equation (1.2). If it was possible to get analytical solution of equation (1.2), one could easily find by putting $\zeta = 1$, that system does not oscillate. However, with the method of numerical techniques, one has to run the program for different values of ζ and then find out that for $\zeta \geq 1$, system is stable. Same is the case with simulation. One has to run simulation model number of time, to arrive at a correct output. Due to this reason, sometimes numerical computation is called *simulation.* But information derived numerically does not constitute simulation. Numerical computation only gives variations of one parameter in terms of other and does not give complete scenario with the time.

Simulation has long been an important tool of designers, whether they are simulating a supersonic jet, a telephone communication system, a wind tunnel, a large scale military war gaming, or a maintenance operation.

Although simulation is often viewed as a "method of last resort" to be employed when every other technique has failed. Recent advances in simulation methodologies, availability of softwares, and technical developments have made simulation one of the most widely used and accepted tools in system analysis and operation research.

Naylor et al., [41] defines the simulation as follows:

Simulation is a numerical technique for conducting experiments on a digital computer, which involves certain types of mathematical and logical models over extended period of real time.

We thus define *system simulation as the technique of solving problems by the observation of the performance, over time, of a dynamic model of the system*. In other words, we can define simulation as an experiment of physical scenario on the computer.

Why simulation is required? According to Naylor [41], some of the reasons why simulation is appropriate are:

1. Simulation makes it possible to study and experiment with the complex internal interactions of a given system, whether it be a firm, an industry, an economy, or some subsystem of one of these.
2. Through simulation we can study the effect of certain informational, organizational, and environmental change on the operation of a system by making alterations in the model of the system and observing the effects of these alterations on the system's behavior.
3. Detailed observation of the system being simulated may lead to a better understanding of the system and to suggestion for improving it, suggestions that otherwise would not be apparent.

4. Simulation can be used as a pedagogical device for teaching both students and practitioners basic skills in theoretical analysis, statistical analysis, and decision-making.
5. Operational gaming has been found to be an excellent means of simulating interest and understanding on the part of the participants, and is particularly useful in the orientation of persons who are experienced in the subject of the game.
6. Simulations of complex systems can yield valuable insight into which variables are more important than others in the system and how these variables interact.
7. Simulation can be used to experiment with new situations about which we have little or no information so as to prepare for what may happen.
8. Simulation can serve as a "pre service test" to try out new policies and decision rules for operating a system, before running the risk of experimenting of the real system.
9. When new components are introduced into a system, simulation can be used to help foresee bottlenecks and other problems that may arise in the operation of the system.

Monte Carlo method of simulation is one of the most powerful techniques of simulation and is discussed below.

1.5.1 Monte Carlo Simulation

Simulation can also be defined as a technique of performing sampling experiments on the model of the system. This is called stochastic simulation and is a part of simulation techniques. Because sampling from a particular probability distribution involves the use of random numbers, stochastic simulation is sometimes called *Monte Carlo Simulation.* Historically, Monte Carlo method is considered to be a technique, using random or pseudo random numbers. It is important to know what random numbers are. Let us take a simple example of tossing a coin. If coin is unbiased, probability of coming head is 0.5. If we generate two numbers say, 0 and 1, so that occurrence of both is equally likely. Let us assume that number 1 depicts head and 0, tail. These numbers are called uniform random numbers. We will discuss stochastic simulation in chapter four.

We give below some differences between the Monte Carlo method and simulation:

1. In the Monte Carlo method, time does not play as substantial role, a role as it does in stochastic simulation.
2. The observations in the Monte Carlo method, as a rule, are independent. In simulation, however, we experiment with the model over time so, as a rule, the observations are serially correlated.
3. In the Monte Carlo method, it is possible to express the response as a rather simple function of the stochastic input variates. In simulation the response is usually a very complicated one and can be expressed explicitly only by the computer program itself.

EXERCISE

1. What is the difference between static and dynamic models?
2. Give an example of a dynamic mathematical model.
3. In the automobile wheel suspension system, it is found that the shock absorber damping force is not strictly proportional to the velocity of the wheel. There is an additional force component equal to D_2 times the acceleration of the wheel. Find the new conditions for ensuring that there are no oscillations.
4. What are the basic steps to be followed while making a model?
5. (*a*) What are distributed lagged models?

   (*b*) If demand and supply of a product obey following equations.

   $D = a + bP$

   $S = c - dP$

   $Y = S$

Here a, b, c, and d are given numbers, convert this model to distributed lagged model (4).

2 PROBABILITY AS USED IN SIMULATION

Events, whose outcome can not be predicted with certainty, are called probabilistic events. Knowledge of probability theory is a prerequisite for the simulation of probabilistic events. Unlike the scientific experiments in engineering, where outcome of results is known, in case of random events it can not be exactly predicted. Outcome of such events generally follow a special pattern which can be expressed in mathematical form called probability distribution. By probability, we mean chances of occurrence of an event. In the present chapter, basic concepts of probability theory are discussed in details for brushing up the knowledge of the students. In nature, number of events are known to be random and results for such events can not be predicted with certainty. Let us take an example of tossing of a coin. No one can tell whether head or tail will be the outcome. But if we toss coin for a large number of time, half of the time, head will be outcome and rest half of the time, tail will be the outcome. In the language of the probability, we say that the probability of coming head on top is 0.5. Similarly failure of a machine, life time of an electric bulb or impact point of a bullet on the target also follows some probability distribution. Estimation of weapon delivery error is one of the important parts of weapon performance studies. When a weapon is launched on a target from a distance, there are chances that it may or may not hit at a desired point. This is due to the inherent error in weapon release mechanism. A simple example will help to understand weapon delivery error. If one tries to drop a stone from a height on a small circle drawn on the ground, it is quite possible that stone may or may not fall inside the circle. Almost same is the problem when a weapon is dropped on a target from a distance. Thus falling of a weapon on some other point than its *aim point* is called *weapon delivery error.* Similarly failure of a machine or one of its parts can not be predicted. In this chapter, we will study that all the phenomena described above can be predicted, but with some probability. The problem in this

Johann Carl Friedrich Gauss

(30 April 1777 – 23 February 1855) was a German mathematician and scientist who contributed significantly to many fields, including number theory, statistics, analysis, differential geometry, geodesy, electrostatics, astronomy, and optics. He made his first groundbreaking mathematical discoveries while still a teenager. He completed *Disquisitiones Arithmeticae*, his magnum opus, in 1798 at the age of 21, though it would not be published until 1801. This work was fundamental in consolidating number theory as a discipline and has shaped the field to the present day.

connection is to evaluate the characteristics of a system in probabilistic or statistical terms and evaluate its effectiveness. In section 2.4, first basic concept about various distribution functions will be given, and then some case studies will be taken up for illustration purpose. First of all knowledge of some basic parameters is essential so as to understand the various probability density functions. These parameters will be discussed in sections from 2.1 to 2.3.

2.1 Basic Probability Concepts

What do we mean if we say, probability of occurrence of an event is 0.5? What is an event? Let us consider a simple experiment of tossing a coin. Can one predict, whether head or tail will come. No, it is not possible to predict. In order to know the answer, we have to toss the coin at least hundred times. Suppose we get 49 times head and 51 times tail. Then probability of getting head in hundred trials is defined as "total number of heads/total number of toss". Thus getting head in tossing a coin is an event. This can be defined in technical language as,

Probability of occurrence of an event = total favourable events/total number of experiments.

In this case probability is 0.49, if all the events are equally likely. This is classical definition of probability. Here equally likely is the condition. In case of coin, it has to be unbiased coin. Now if we toss the same coin more number of times say 1000 times, we will come to know that probability of getting head or tail in case of unbiased coin is closer to 0.5. This approach to probability is *empirical approach.* Higher is the number of trials, accurate will be the probability of success.

In probability, all favourable and unfavourable events are called *sample points.* In above example of a coin, if we toss a coin thousand times, then thousand outcomes are sample points. Thus we define;

2.1.1 Sample Point

Each possible outcome of an experiment is called a *sample point.*

2.1.2 Sample Space

The *sample space* is the set of all possible sample points of an experiment.

To understand these definitions, let us take an example of a *dice*. If we throw this *dice*, any outcome (number on the top side of a dice) of it will be a sample point. And the sample space will be a set of all the outcomes taken together i.e., S {1, 2, 3, 4, 5, 6} will be a sample space, where symbol { } defines a set of sample points 1, 2, 3, 4, 5, 6. If sample space S has finite number of points, it is called *finite sample space*. Now let us consider a set of all the natural numbers. Can you count these numbers? Of course one can count but up to what number. There are infinite natural numbers. You surely can not count up to infinity, but yet you can count up to your capacity. Thus a sample space consisting of all the natural numbers 1, 2, 3, 4, 5,... is called a *countable infinite sample space*. It is infinite yet countable. Now let us consider a different case i.e., number of points on a number line. If sample space S has as many points as there are numbers in some interval (0,1) on line, such as $0 \leq x \leq 1$, it is called a *non-countable infinite sample space*. There are infinite points on a line of unit length, but can you count all these points. Of course, not. A sample space that is finite or countable infinite is called a *discrete sample space*, while one that is non-countable infinite is called *non-discrete (continuous) sample space*.

2.1.3 Event

Let us define another parameter called event. An event is a subset of a sample space S, i.e., it is a set of possible outcomes. If an outcome of an experiment is an element of subset A, we say the event A has occurred. An event consisting of single point of S is often called a *simple event*.

We have used above a word set. What is a set? A set can be defined as a collection of similar types of items. For example, outcome of throw of a pair of dice is a set. This set is nothing but numbers from 2 to 12.

If we express the outcomes of an experiment in terms of a numerically valued variable X, which can assume only a finite or denumerable number of values, each with a certain probability (By denumerable we mean that values can be put into a one to one correspondence with the positive integers.), then such a variable is called a *random variable* (also called *stochastic variable or variate*) and will fall in a sample space. **Here onward a random variable will be denoted by a capital letters while the values of random variables will be denoted by small letters**. Let us consider an example of a random variable. For example, if we roll a pair of dice, sum X of two numbers which turn up, must be an integer between 2 and 12. But it is not possible to predict which value of X will occur in the next trial. If we can predict the chance of a particular number to come in the next trial then such an outcome of trial is called probability of occurrence of that number which will be the value of the event, called *random variable.* Therefore we can say that, if X depends on chance (a probability is assigned to random variable X) and a probability can be attached to it then it is a random variable. Or it can be predicted that next time when a dice is rolled, what will be the outcome.

2.1.4 Universal Set

In general, a sample space is said to be discrete if it has finitely many or countable infinite elements. A sample space, which is called *universal set,* can have number of subsets. For example, in the example of throw of a pair of dice, set of all even outcome can be called *one subset* and that of odd outcomes can be called *second subset.*

2.1.5 Set Operations

Theory of sets is a field in itself and we will just give a brief introduction here. By using set operations on events in *S,* we can obtain few other events in *S.* For example if A and B are events, then

1. $A \cup B$ is the event, "either A or B or both", is called *union* of A and B.
2. $A \cap B$ is the event, "both A and B ", is called *intersection* of A and B.
3. A^c is the event "not A" and is called complement of A and is equal to $(S - A)$.
4. $A - B = A \cap B^c$ is the event "A but not B".

In order to further elaborate the above definitions, below we define three subsets of a sample space as,

$$A = \{1, 2, 3\},\ B = \{3, 2, 5, 6, 7\} \text{ and } C = \{a, b, c\}$$

Then the union, intersection and complement of sets are,

(*a*) $S = \{1, 2, 3, 5, 6, 7, a, b, c\}$

(*b*) $A \cup B = \{1, 2, 3, 5, 6, 7\}$

(*c*) $A \cap B = \{2, 3\}$

(*d*) $A^c = (5, 6, 7, a, b, c\}$

Using concept of Venn diagram, above relations can be shown as,

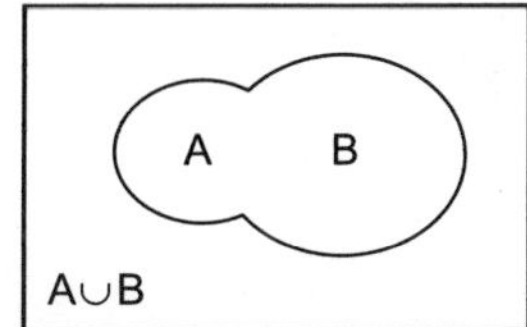

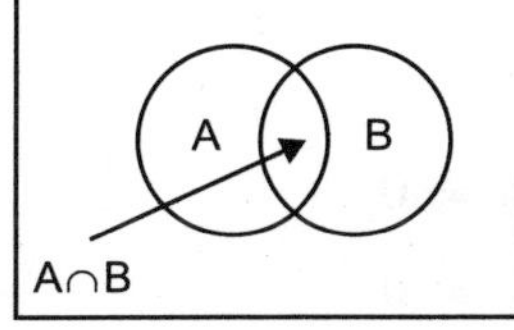

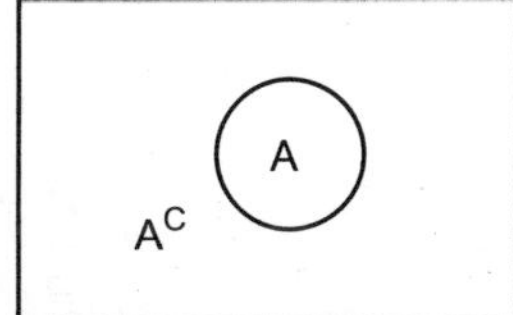

Fig. 2.1: *Venn diagram for union, intersection and complement of three sets.*

Concept of union and intersection will be used frequently in coming sections.

2.1.6 Statistical Independence

We now give definition of statistically independent random events. Two random events A and B are statistically independent if and only if

$$P(A \cap B) = P(A)P(B) \quad ...(2.1)$$

Thus, if A and B are independent, then their joint probability can be expressed as a simple product of their individual probabilities.

Equivalently, for two independent events A and B,

$$P(A|B) = P(A)$$

and

$$P(B|A) = P(B).$$

Symbol $P(A|B)$ here means, probability of occurrence of A if B has already occurred. In other words, if A and B are independent, then the conditional probability of A, given B is simply the individual probability of A alone; likewise, the probability of B given A is simply the probability of B alone. This result is after Swine and is called SWINE'S THEOREM.

We further elaborate the concept to understand this. If X is a real-valued random variable and a is a number, then the event that $X \le a$ is an event, so it makes sense to speak of its being, or not being, independent of another event.

Two random variables X and Y are independent if and only if, for any numbers a and b the events $[X \le a]$ (the event of X being less than or equal to a) and $[Y \le b]$ are independent events as defined above. Similarly an arbitrary collection of random variables—possible more than just two of them—is independent precisely if for any finite collection $X_1, ..., X_n$ and any finite set of numbers $a_1, ..., a_n$, the events $[X_1 \le a_1], ..., [X_n \le a_n]$ are independent events as defined above.

2.1.7 Mutual Exclusivity

In section 2.1.6, we define intersection of event A and B as

$$P(A \cap B) = P(A)P(B|A)$$

This equation is read as "Probability of occurrence of intersection of event A and B is equal to the products of probability of occurrence of A and probability of occurrence of B if A has already occurred." Now two events A and B are defined as *mutually exclusive* if and only if

$$P(A \cap B) = 0$$

as long as $P(A \cap B) \neq 0$

and $P(B) \neq 0$

Then $P(A|B) \neq 0$

and $P(B|A) \neq 0$

In other words, the probability of A happening, given that B happens, is nil since A and B cannot both happen in the same situation; likewise, the probability of B happening, given that A happens, is also nil.

2.1.8 The Axioms of Probabilities

If sample space S is discrete, all its subsets correspond to events and conversely, but if S is not discrete, only special subsets (called measurable) correspond to events. To each event A in the class C of events (events by throw of dice is one class where as toss of coin is another class), a real number $P(A)$ is associated. Then P is called a **probability function**, and $P(A)$ the probability of the event A, if the following axioms are satisfied.

Axiom 1: For every event A in the class C, $P(A) \geq 0$.

Axiom 2: For the sure or certain S in the class C, $P(S) = 1$.

Axiom 3: For any number of mutually exclusive events $A_1, A_2, \ldots$ in the class C,

$$P(A_1 \cup A_2 \cup \ldots) = P(A_1) + P(A_2) + \cdots$$

Below some useful rules of probability, derived from the basic rules and illustrated by Venn diagrams, where probability is defined by relative areas has been given.

From Venn diagram, it is clear that probability of occurrence A and B together is sum of the probabilities minus the common area, which is defined as follows,

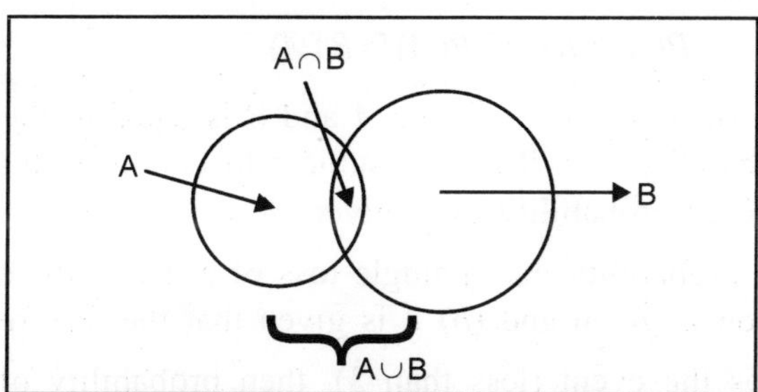

$$P(A \cup B) = P(A) + P(B) - P(A \cap B)$$

This is the case when A and B are not mutually exclusive ($A \cap B \neq 0$).

(*a*) **Complement Rules:** The probability of the complement of A is

$$P(\text{not } A) = P(A^c) = 1 - P(A)$$

Proof: Since space S is partitioned into A and A^c, and $P(S) = 1$ (Fig. 2.2),

$$1 = P(A) + P(A^c)$$

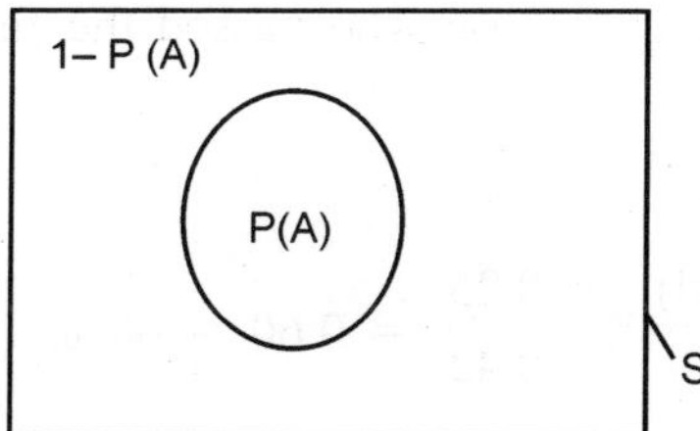

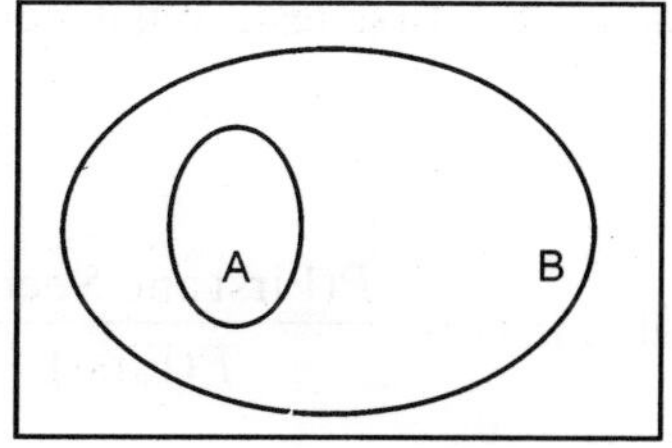

Fig. 2.2: *Venn diagram for complement and difference rule.*

(*b*) **Difference Rule:** If occurrence of A implies occurrence of B, then $P(A) \leq P(B)$, and the difference between these probabilities is the probability that B occurs and A does not:

$$P(B \text{ and not } A) = P(BA^c) = P(B)\ (1-P(A))$$

Proof: Since every outcome in A is an outcome in B therefore A is a subset of B. Since B can be partitioned into A and (B but not A),

$$P(B) = P(A) + P(BA^c)$$

hence the result.

Now since probability of an outcome A in one trial is $P(A)$ then probability of not occurrence of A is $1 - P(A)$. Probability of not occurrence in two trials is that it does not occur in first trial and it also does not occur in second trial, which is same as $(1 - P(A))^2$. Similarly if we extrapolate, probability of not occurrence of A in n trials is $(1 - P(A))^n$. This means A does not occur in first trial and it also does not occur in all the n trials. Therefore probability that A occurs at least once in n trials is $1 - (1 - P(A))^n$. This result is very useful and will be used in next chapters.

2.1.9 Conditional Probabilities

Let A and B be the two events such that $P(A) > 0$. Denote by $P(B|A)$ the probability of B given that A has occurred. Since A has already occurred, it becomes the new sample space replacing the original S. From this we come to the definition

$$P(A \cap B) \equiv P(A)P(B \mid A) \qquad ...(2.2)$$

This means, probability of occurrence of both A and B is equal to the product of the probability that A occurs time the probability that B occurs subject to condition that A has already occurred. We call $P(B|A)$ the conditional probability of B given A.

Example 2.1: Find the probability that a single toss of a dice will result in a number less than 3, if (*a*) no other information is given and (*b*) it is given that the toss resulted in an even number.

Solution: Let B denotes the event (less than 3), then probability of occurrence of B is,

(*a*) $P(B) = P(1) + P(2) = 1/6 + 1/6 = 1/3$ ...(2.3)

Assuming that all the occurrence are equally likely.

(*b*) Let A be the event (even number), then $P(A) = 3/6 = 1/2$. Also $P(A \cap B) = 2/6 = 1/3$. Then

$$P(B|A) = \frac{P(A \cap B)}{P(A)} = \frac{1/3}{1/2} = 2/3$$

which is the probability of occurrence of B when A has already occurred.

Example 2.2: A maths teacher gave her class two tests. 25% of the class passed both tests and 42% of the class passed the first test. What percent of those who passed the first test also passed the second test?

Solution:

$$P\,(\text{Second} \mid \text{First}) = \frac{P(\text{First and Second})}{P(\text{First})} = \frac{0.25}{0.42} = 0.60 = 60\%.$$

2.2 Discrete Random Variable

Random events have been studied in section 2.1. Here we give definition of *random variable.* In most practical cases the random variables are either discrete or continuous. In the present section discrete random variables will be discussed. A discrete random variable is defined as:

A random variable X and its corresponding distribution are said to be discrete, if X has the following properties.

(1) The number of values for which X has a probability different from 0, is finite or utmost countable infinite.

(2) Each finite interval on the real line contains at most finitely many of those values. If an interval $a \le X \le b$ does not contain such a value, then $P\ (a < X \le b) = 0$. Here a and b are the upper and lower limits of the stochastic variable X.

If x_1, x_2, x_3 be values for which X has positive corresponding probabilities p_1, p_2, p_3, then function

$$\begin{aligned} f(x) &= \Pr\ (X = x_j) = p_j,\ \ j = 1, 2, ..., n \\ &= 0, \text{ otherwise} \end{aligned} \quad ...(2.4)$$

with the condition,

$$\left.\begin{aligned} &(i)\ f(x) \ge 0 \\ &(ii)\ \sum_x f(x) = 1 \end{aligned}\right\} \quad ...(2.4a)$$

Here $f(x)$ is called the **Probability Density Function** (PDF) of X and determines the distribution of the random variable X. In the second equation of (2.4*a*) summation is over the all possible values of x. Meaning of expression $\Pr(X = x_j) = p_j$ here, is that the probability that $X = x_j$ is p_j. Equation (2.4) says that distribution of random variable X is given by function $f(x)$ which is equal to p_j for $x = p_j$. To understand this, let us take an example of two dice. In this example, if we throw two dice together, the probability of getting a sum two of the two faces is 1/36 (There are total 36 ways, when two dice can fall, and sum two can come only in one way i.e., (1 + 1 = 2). Thus we can say,

$$f(2) = 1/36$$

similarly sum four can come in three ways (1 + 3, 3 + 1, 2 + 2), thus

$$f(4) = 3/36$$

and so on. Many times the PDF is in the form of a table of probabilities associated with the values of the corresponding random variable whereas sometimes it might be expressed in some closed form, such as

$$f(x) = \Pr(X = x) = p^x (1-p)^{n-x};\ x = 0, 1, ..., n \quad \text{and} \quad 0 < p < 1 \quad ...(2.5)$$

Equation (2.5) will be discussed in details in section 2.4.1. At this stage, we will assume that $f(x)$ is some analytic function. In the coming sections, we will discuss number of probability density functions which are in terms of closed form functions.

2.3 Expected Value and Variance of a Discrete Random Variable

A quantity called Expected Value (EV), which is associated with every random variable will be discussed in this section. Expected value, also called mean of the discrete random data denoted by μ, is the sum of the product of all the values, a random variable takes, with its probabilities assigned at those values. Thus expected value is defined as [23]

$$E(X) = \Sigma_{R(x)} \Pr(X = x)\cdot x$$
$$= \Sigma_{R(x)} f(x)\cdot x \qquad ...(2.6)$$

where $R(x)$ denotes the range of X (Sample Space) and Σ is a symbol of summation which means, sum of all the values over the sample space $R(x)$. Range or sample space R is the region in which all the values x of stochastic variable X fall. We can call expected value as the mean of values of X in the range R. This will be clear from the exercise 2.3 given below.

2.3.1 Some Theorems on Expected Value

Expected value of a probability density function has following properties,

(*a*) If c is any constant, then

$$E(cX) = cE(X) \qquad ...(2.6a)$$

(*b*) If X and Y are two independent random variables, then

$$E(X + Y) = E(X) + E(Y) \qquad ...(2.6b)$$

(*c*) If X and Y are two independent random variables, then

$$E(XY) = E(X)\cdot E(Y) \qquad ...(2.6c)$$

Example 2.3: In the Table 2.1, probability of arrival of sum of the numbers on two faces of two dice thrown simultaneously, are given. Calculate the value of $E(x)$?

Table 2.1: Probability of occurrence of number (X) in a throw of two dice

Number expected (X)	Probability f(x)	Number expected (X)	Probability f(x)
2	$\frac{1}{36}$	7	$\frac{6}{36}$
3	$\frac{2}{36}$	8	$\frac{5}{36}$
4	$\frac{3}{36}$	9	$\frac{4}{36}$
5	$\frac{4}{36}$	10	$\frac{3}{36}$
6	$\frac{5}{36}$	11	$\frac{2}{36}$
		12	$\frac{1}{36}$

Solution: It can be seen from the Table 2.1 that,

$$E(x) = \sum f(x)\cdot x$$
$$= (1/36)\cdot[2 + 2\times 3 + 3\times 4 + 4\times 5 + 5\times 6 + 6\times 7 + 5\times 8 + 4\times 9 + 3\times 10 + 2\times 11 + 12]$$
$$= 7$$

which is nothing but the mean of the values of X (Fig. 2.3).

2.3.2 Variance

Another important quantity closely associated with every random variable is its *variance*. Basically, the variance is a measure of the relative spread in the values, the random variable takes on. In particular, the variance of a discrete random variable is defined as

$$\text{Var}(X) = E\,\{X - E(X)\}^2$$

$$= \Sigma_{R(X)} \Pr(X = x)[x - E(x)]^2$$

$$= \Sigma_{R(X)} f(x)[x - E(x)]^2 \qquad \text{...(2.6d)}$$

This will be clear when we work out the exercise 2.4.

2.3.3 Some Theorems on Variance

We define here another important parameter called *standard deviation* and is denoted by symbol σ. Standard deviation is nothing but square root of variance i.e.,

(*a*) $\sigma^2 = E[(X - \mu)^2] = E(X^2) - \mu^2 = E(X^2) - [E(X)]^2$ ...(2.6*e*)
where $\mu = E(X)$, and σ is called standard deviation.

(*b*) If c is any constant,
$\text{Var}\ (cX) = c^2\, \text{Var}\ (X)$ ...(2.6*f*)

(*c*) The quantity $E\,[(X - a)^2]$ is minimum when $a = \mu = E(X)$ ...(2.6*g*)
If X and Y are independent variables,

(*d*) $\text{Var}(X+Y) = \text{Var}(X) + \text{Var}(Y)$ or $\sigma^2_{X+Y} = \sigma^2_X + \sigma^2_Y$
$\text{Var}(X-Y) = \text{Var}(X) + \text{Var}(Y)$ or $\sigma^2_{X-Y} = \sigma^2_X + \sigma^2_Y$ ...(2.6*h*)

Generalisation of these laws can always be made for more number of variables.

Example 2.4: Determine the variance (Var) of the random variables X given in the Table 2.1.

Solution:

$$\begin{aligned}\text{Var}\ (x) &= \Sigma f(x)\cdot[x - E(x)]^2 \\ &= 1/36[\ 1(-5)^2 + 2(-4)^2 + 3(-3)^2 + 4(-2)^2 + 5(-1)^2 + 6(0)^2 \\ &\qquad + 5(1)^2 + 4(2)^2 + 3(3)^2 + 2(4)^2 + 1(5)^2] \\ &= 1/36[25 + 32 + 27 + 16 + 5 + 0 + 5 + 16 + 27 + 32 + 25] \\ &= 210/36 \\ &= 5.83333\end{aligned}$$

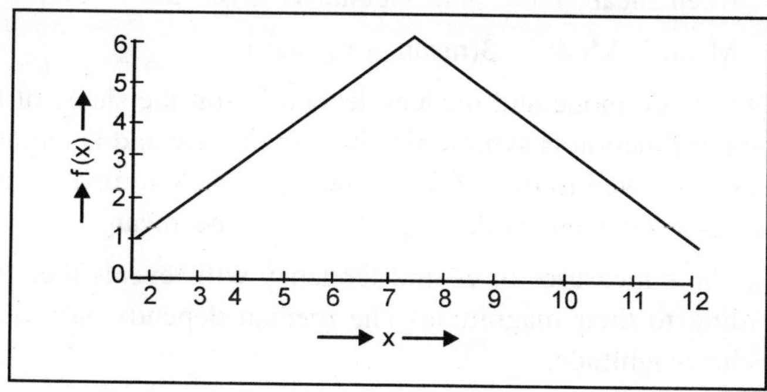

Fig. 2.3: *Probability of occurrence of number in a throw of two dice.*

In Fig. 2.3, variation of $f(x)$ versus x has been shown. Variation of $f(x)$ is of the type of an isosceles triangle, vertex being at $x = 7$, $f(x) = 5.83333$. This means that $x = 7$ is the mean value of the variables x for which maximum spread in $f(x)$ is 5.83333. Relation (2.6*e*) shows, Var(x) is nothing but sum of the squares of deviation of x from its mean value $E(x)$ multiplied by the probability density function $f(x)$ at x. Square root of Var(x) is called the standard deviation of the variable X and is denoted by symbol σ.

2.4 MEASURE OF PROBABILITY FUNCTION

Before we discuss various probability distribution functions, it is important to define two important characteristics of a probability density function i.e., central tendency and dispersion.

2.4.1 Central Tendency

The three important measures of central tendency are mean, mode and median. Mean has already been defined (section 2.3). Mode is the measure of peak of distribution. Value of x at which probability distribution function $f(x)$ is maximum is called mode of the distribution. A distribution can have more than one mode. Such a distribution is called *multimodal.* In case of discrete distribution, mode is determined by the following inequalities,

$$p(x = x_i) \leq p(x = \hat{x}), \text{ where } x_i \leq \hat{x}$$

and for a continuous distribution $f(x)$, mode is determined as,

$$\frac{d}{dx}[f(X)] = 0$$

$$\frac{d^2}{dx^2}[f(X)] < 0$$

which is the condition that $f(x)$ is maximum.

2.4.2 Median

Median divides the observations of the variable in two equal parts. Thus for a discrete or continuous distribution of variable x, if X denotes the median, then

$$p(x \leq X) = p(x \geq X) = 0.5$$

Median can easily be found from Cumulative Distribution Function because at median value of CDF = 0.5. Relation between mean, mode and median is given as,

$$\text{Mean} - \text{Mode} = 3(\text{mean} - \text{median}).$$

The relative values of mean, mode and median depends upon the shape of the probability curve. If the probability distribution function is symmetric about the centre and is unimodal then all the three will coincide. If the curve is skewed to the left as in the Fig. 2.4 then mode is greater than the mean, but if it is skewed towards right then mode is greater than the mean.

A comparison of the three measures of central tendency will reveals that, the mean involves the weightage of data according to their magnitude. The median depends only on the order, while the mode depends only on the magnitude.

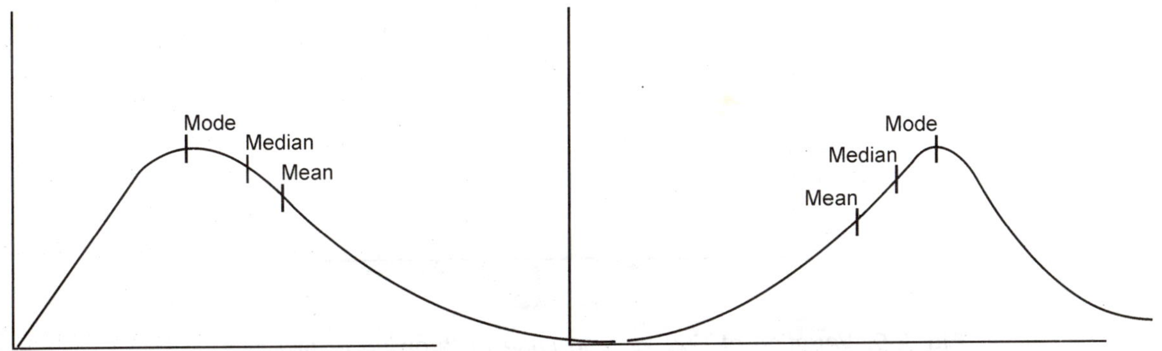

Fig. 2.4: *Demonstration of mean, mode and median.*

2.5 Some Important Distribution Functions

Quite often a problem can be formulated such that one or more of the random variables involved will have a Probability Density Function (PDF). It is possible to put this function in a closed form. There are number of probability density functions, which are used in various physical problems. In this section we will examine some of the well known probability density functions. In the previous section, it has been shown that probability density function of the outcomes of throwing the two dice follows a distribution which is like an isosceles triangle, and mean of all these of variables is 7. From the inspection of Table 2.1, we can easily see that in this case, probability density function can be written in an analytic form as

$$f(x) = \begin{cases} \dfrac{x-1}{36}, 2 \le x \le 7 \\ \dfrac{13-x}{36}, 7 \le x \le 12 \end{cases}$$

Here we have very easily expressed PDF of two dice in the form of an analytic equation. But this may not always be so simple. In practice we often face problems in which one or two random variables follow some distribution which is well known or it can be written in analytic form. In the next section, we will discuss few common distribution functions which we will need for the development of various models.

2.5.1 Cumulative Distribution Function

Closely related to Probability Distribution Function (PDF) is a function called **Cumulative Distribution Function** (PDF). This function is denoted by $F(x)$. Cumulative distribution function of a random variable X is defined as

$$F(x) = \Pr(X \le x)$$

which can be expressed as,

$$F(x) = \sum_{X \le x} f(z) \qquad ...(2.7)$$

$F(x)$ represents a probability that the random variable X takes on a value less than or equal to x. We can see that in the example of rolling pair of dice only two outcomes less than or equal to three are $f(2)$ and $f(3)$, thus (Example 2.1).

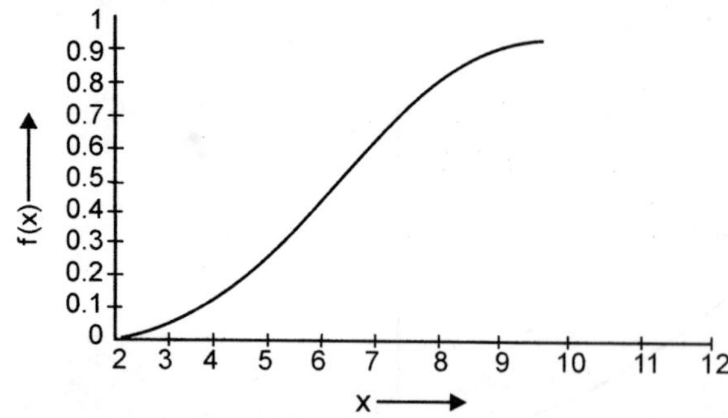

Fig. 2.5: *Variation of F(x) vs. x in case of throwing of two dice.*

$$F(3) = \Pr(X \leq 3) = f(2) + f(3)$$

$$= \frac{1}{36} + \frac{2}{36} = \frac{3}{36}$$

From Fig. 2.5, we see that function $F(x)$ increases with the increase of parameter x, whereas in the case of function $f(x)$, this rule is not compulsory rule. It can be seen from Table 2.1 that sum of probabilities of all the events is unity. That is

$$\sum_{R(X)} p(x_i) = 1$$

which is nothing but proof of axiom 2 of definition of probability (section 2.1.8).

2.5.2 Uniform Distribution Function

The uniform distribution has wide application in various problems in modeling. Flow of traffic on road, distribution of personnel in battle field, and distribution of stars in sky are examples of uniform distribution. It is the basic distribution required for calculating the other distributions. For the evaluation of different types of warheads, it is generally assumed that the distribution of ground targets (say distribution of personnel and vehicles in a battle field) is randomly uniform. Even most of the warheads have sub-munitions with uniform ground patterns.

A random variable X which is uniformly distributed in an interval $[a, b]$ can be defined as

$$f(x) = \frac{1}{b-a}, \text{ for } a \leq x \leq b \qquad \text{...(2.8)}$$

The mean and standard deviation of uniform distribution is given by

$$\mu = \frac{a+b}{2}, \quad \sigma^2 = \int_a^b (x-\mu)^2 f(x)dx$$

$$= \frac{1}{b-a}\int_a^b \left(x - \frac{a+b}{2}\right)^2 dx$$

$$= \frac{1}{3(b-a)}\left[\left(x - \frac{a+b}{2}\right)^3\right]_a^b$$

$$= \frac{1}{3(b-a)}\left[\left(\frac{b-a}{2}\right)^3 - \left(\frac{a-b}{2}\right)^3\right]$$

$$= \frac{(b-a)^2}{12} \quad ...(2.8a)$$

Cumulative distribution function of uniform distribution is

$$F(x) = \int_a^x \frac{1}{b-a}\, dx = \frac{x-a}{b-a} \quad ...(2.8b)$$

The application of uniform distribution will be discussed in chapter 4.

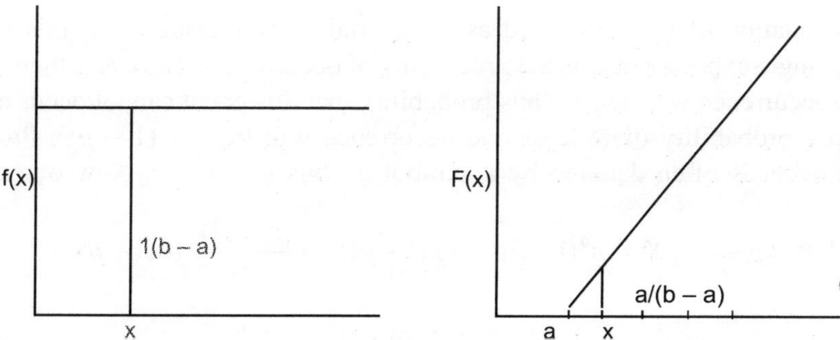

Fig. 2.6: *PDF and CDF for uniform distribution.*

In Fig. 2.6 we have given variation of $f(x)$ and $F(x)$. It can be seen that $f(x)$ is a horizontal straight line whereas $F(x)$ is an inclined line. As x is increased by a, $F(x)$ is incremented by $a/(b-a)$.

2.5.3 Binomial Distribution Function

Suppose an experiment can yield only two possible outcomes, say 0 or 1, where 0 represents a failure and 1 represents a success. For example on tossing a coin we get either a head or a tail. If head is the outcome of tossing, we say experiment is a success i.e., $p = 1$, otherwise it is a failure ($p = 0$), where p is the probability of a success in each trial of an experiment. Now repeat the trial n times under identical conditions. If x represent the number of successes in the n experiments, then x is said to have a binomial distribution whose PDF is given by:

$$f(x) = \Pr(X = x) = C_x^n p^x (1-p)^{n-x} \quad ...(2.9)$$

where, $x = 0,1,..., n;\quad 0 < p < 1$

Here C_x^n is defined as number of ways in which there are x successes in n trials. C_x^n can also be defined as combination of n items taken x at a time. To understand C_x^n we consider an example. If we have three numbers 1, 2, and 3, and want to make combination of these numbers taken two at a time, these combinations will be,

12, 13 and 23

This can be written as

$$C_2^3 = 3$$

Thus combination of three numbers taken two at a time is three. We have assumed in the definition of C_x^n that combinations 23 and 32 are same. Thus C_x^n denotes the number of ways k success can occur in n independent experiments or trials and is given by

$$C_x^n = \frac{n!}{x!(n-x)!} = \frac{n(n-1)(n-2)(n-3)\cdots 1}{x(x-1)(x-2)\cdots 2\cdot 1\cdot (n-x)\cdots(n-k-1)\cdots 1} \quad ...(2.10)$$

where

$$n! = n(n-1)\cdot(n-2)\cdots 3.2.1 \quad ...(2.10a)$$

In the above equation $n!$ is pronounced as n factorial. To understand binomial distribution, we assume that if in a single experiment, p is the probability of occurrence of an event then in n experiments probability of it's occurrence will be p^n. Thus probability that this event cannot occur in n experiments will be $(1-p)^n$ and probability of at least one occurrence will be $1-(1-p)^n$. Probability of non occurrence of an event is often denoted by a symbol q, thus $q = 1 - p$. Now one can easily write

$$1 = (p+1-p)^n = p^0(1-p)^n + np(1-p)^{n-1} + \frac{n(n-1)}{2!}p^2(1-p)^{n-2}$$

$$+\cdots+np^{n-1}(1+P)+P^n(1-P)^0 \quad ...(2.11)$$

This equation has been obtained by binomial expansion. For Binomial expansion, readers may refer to some book on Differential calculus [30]. We can express equation (2.11) in a closed form as,

$$\sum_{x=0}^{n} C_x^n\, p^x(1-p)^{n-x} = 1 \quad ...(2.12)$$

Thus we define a binomial distribution function as

$$f(x) = C_x^n p^x(1-p)^{n-x} \quad ...(2.13)$$

Expected value and variation of binomial distribution function is given by [24,72] (Appendix-2.1)

$$E(X) = np$$

$$\text{Var}(X) = np(1-p) \quad ...(2.14)$$

Let us understand binomial distribution function by following example.

Example 2.5: In an examination, total thirty students appeared. If probability of passing the examination of one student is 0.6 (i.e., $p = 0.6$), then what is the probability that none of the student will pass and twenty students will pass.

Solution:

This problem can easily be solved by Binomial distribution. Probability of failing all the students will be $f(0)$ and passing of 20 students will be $f(20)$. Thus using equation (2.13), one gets

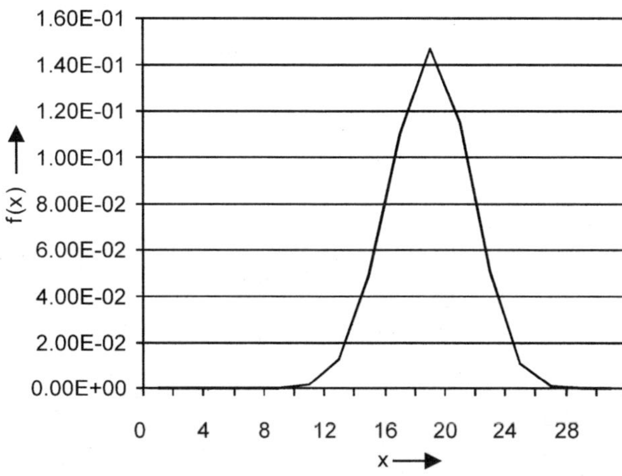

Fig. 2.6a: *Probability of passing of students.*

$$f(0) = C_0^{30}(0.6)^0(0.4)^{30} = 1153 \times 10^{-12}$$

$$f(20) = C_{20}^{30}(0.6)^{20}(0.4)^{10} = 0.115185$$

Table 2.2: Data types used in C++ programming

Type of data	*Range*	*Size in bits*
Signed char	–128 to 127	8
Short int and int	–32768 to 32767	16
Long int	–2147483,648 to 2147483,647	32
Float	–3.4e–38 to 3.4e+38	32
Double	–1.7e–308 to 1.7e+308	64
Long double	–3.4e–4932 to 3.4e+4932	80

This problem has a direct physical relevance. Since probability of failing one student is 0.4, this means that probability of failing all the twenty students is remote. Thus $f(0)$ is a very small number. At the most 12 students are expected to fail and 18 expected to pass ($E(x) = 18$) out of 30 students. In Fig. 2.6*a*, we have drawn probability $f(x)$ vs. x, where x is the number of passing students. This means in another words that probability of passing 18 students is maximum ($f(18) = 0.147375$) in this case. Computer program written in *C++* is given (Program 2.1) for the interest of readers. As has been mentioned in the beginning, knowledge of numerical computation and *C++* programming is a pre-requisite for understanding the contents of this book. *C++* is the most versatile and latest language in programming. We will explain all the steps in *C* programming whenever they are given. However student is advised to refer any book on *C++* programming[3].

Program 2.1: Binomial Distribution Program

```
#include <iostream.h>//Header files to be included. Note that comment command
can be used anywhere.
   #include <math.h>
   double n;
```

```
class comp //defined a class named comp
{
private:
    double x;// Private parameters can not be called outside class.
    double p;// But public function of class can call them.
    double c_n_x;
public:
    void pre_input();
    void input();
    void C_n_x();
    double fact(double);
    void fx();
};
void comp::pre_input
{
    cout<<"n=";
/*cout<< writes on the console "n=" and value of n is to be typed here.Symbol
<< means give to (console) */
      cin>>n; //cin>> will read the value of n typed on the console.
      cout<<"p=";
      cin>>p;
}
void comp::input()// input() is a function defined in the class comp and
{
```

Note-1 on C++: A line starting with // in a program is comment and computer will not execute this comment. C++ compiler has some built in libraries, which one has to include in the beginning of program to execute some built in functions. For example iostream.h in program 2.1 is basic input output stream file and .h denotes it is a header file. Another file which is included in this program is math.h. Whenever program uses some mathematical functions, math.h has to be included. Next parameter is defined as **class**. A class is nothing but definition of all the parameters and functions used in the program. It has two types of parameter definitions, private and public. What exactly functions do, we have explained in the program, whereever required using comment command //. Anything written in between /*....*/ is also treated as comments. This is the command of C language but is valid in C++ also.

Note-2 on C++: In the program 2.1, we have explained all the commands of C++ wherever required. Term **double** has not been explained. It is in fact data type definition of a parameter. All the parameters to be used in C++ have to be defined in the program. In C language these parameters are required to be defined in the main program in the beginning. But in C++, these parameters can be defined any where in the function, where first time they are used. Parameter can be integer (**int**), fraction (to be defined as **float**). In table 2.2 we give the size of numbers to be stored in computer along with their definitions. We have used so far double in first C++ program. Other data types are also put in the table for future reference purpose.

```
        cout<<"x="; //reads value of x as above in pre-input()
        cin>>x;
    }
    void comp::C_n_x() // Function C_n_x() computes value of C^n_x.
    {
        double tmp1 = fact(n);
    //Calls the function fact(n), which computes factorial of n. Each line
//in C++ ends with ;
        double tmp2 = (fact(x)*fact(n-x));
        c_n_x = tmp1/tmp2;
        //c_n_x = C^n_x = fact(n)/(fact(x)*fact(n-x));

        cout<<"c_n_x = "<<c_n_x<<endl;
     // Writes on console the value of C^n_x . Endl means end of line .
    }
    void comp::fx()
    {
    /*When ever we write a function defined in a class we write class name::
function name. Symbol :: is called scope resolution operator. Function fx()
computes Binomial probability distribution function
        f(x) = C^n_x p^x (1-p)^(n-x). */
        double fx;
        fx= c_n_x* (pow(p,x)*(pow((1-p),(n-x))));
    // Function pow(p,x) means p^x.
        cout<<"fx =" <<fx<<endl;
    }
    double comp::fact(double x)
    {
    /*This function computes the factorial of x. First we define fact=1 as initial
value. Function for (int j=x; j > 0; j--) says multiply fact with j and store to
fact in each loop, where j varies from x to 1. j-means decrease the value of j by one
every time control enters the for () loop*/
        double fact = 1.0;
        for (int j = x; j>0; j--)
        {
            fact *=j;
        }
        return fact;//Returns the final value of factorial x.
    }
    void main()
    {
    /* Object obj of class comp has been declared.
```

```
        comp obj;
        obj.pre_input();/* Will call function pre_input()defined in class
comp*/
        for(int i=0;i<=n;++i)
        {
    //for() loop is executed n times.
            obj.input();
            obj.C_n_x();
            obj.fx();
        }}
```

2.5.4 Poisson's Distribution

Another distribution which will be used in combat survivability analysis is Poisson's distribution after SD Poisson[1]. This distribution function is required for the cases, where probability of success of an event is very low in large number of trials. Interestingly, this probability density function was first devised to study the number of cavalry soldiers who died annually due to direct hit by horses on their head. The PDF for a random variable X has the variates

$$f(x) = \Pr(X = x)$$

$$= \frac{e^{-\lambda}\lambda^x}{x!}, \text{ where } x = 0, 1, 2,... \qquad ...(2.15)$$

$$\text{and } \lambda > 0.$$

Here λ is the expected value of the Poisson's distribution. Poisson's distribution is used in the case where out of large number of trials, probability of success of an event is quite low. For example, if a shell explodes in the near vicinity of an aircraft, probability of hitting only two or three fragments to a vital part, out of large number of fragments is too small and is calculated by using Poisson's distribution [54]. It can be proved that sum of all the probabilities when x varies from zero to infinity is unity i.e.,

$$\sum_{x=0}^{n} \frac{e^{-\lambda}\lambda^x}{x!} = 1 \text{ when } n \to \infty$$

Proof:

$$\sum_{x=0}^{n} \frac{e^{-\lambda}\lambda^x}{x!} = e^{-\lambda}\sum_{x=0}^{n} \frac{\lambda^x}{x!} = e^{-\lambda}.e^{\lambda} = 1$$

where

$$\sum_{x=0}^{n} \frac{\lambda^x}{x!} = 1$$

is the Maclaurin series for e^{λ}.

1. *Sime'on Denis Poisson reported this distribution for the first time in his book "Recherches sur la probabilite des judgements (1937)" (Researches on the probability of opinions) [20].*

It can be shown that, if we put $p = \lambda/n$ where n is very large, binomial distribution reduces to Poisson distribution i.e., as n tends to infinity.

Expression for Binomial distribution function is

$$b(x;n,p) = C_x^n P^x (1-p)^{n-x}$$

putting value of $p = \lambda/n$ in this expression, we get

$$b(x;n,p) = C_x^n (\lambda/n)^x (1-\lambda/n)^{n-x}$$

$$= \frac{n(n-1)\cdot(n-2)\cdots(n-x+1)}{n^x \cdot x!} \lambda^x (1-\lambda/n)^{n-x}$$

$$= \frac{\left(1-\frac{1}{n}\right)\cdot\left(1-\frac{2}{n}\right)\cdots\left(1-\frac{x-1}{n}\right)}{x!} \lambda^x (1-\lambda/n)^{n-x}$$

If we now let $n \to \infty$, we get

$$\left(1-\frac{1}{n}\right)\cdot\left(1-\frac{2}{n}\right)\cdots\left(1-\frac{x-1}{n}\right) \to 1$$

and

$$\left(1-\frac{\lambda}{n}\right)^{n-x} = \left[\left(1-\frac{\lambda}{n}\right)^{n/\lambda}\right]^{\lambda}\left(1-\frac{\lambda}{n}\right)^{-x} \to e^{-\lambda}$$

Since $\left[\left(1-\frac{\lambda}{n}\right)^{n/\lambda}\right] = 1/e$ and $\left(1-\frac{\lambda}{n}\right)^{-x} = 1$ when $n \to \infty$

and hence we get the Poisson distribution as

$$= \frac{e^{-\lambda}\lambda^x}{x!}; \quad \begin{matrix} x = 0,1,2,\ldots; \\ \lambda > 0 \end{matrix} \qquad \text{...(2.16)}$$

Expected value and variance for Poisson random variable are

$$E(X) = \lambda$$

$$\text{Var}(X) = \lambda \qquad \text{...(2.17)}$$

Below, we will demonstrate by a numerical example the validity of equation (2.16).

In the Table 2.3, we have given a comparison of both the distributions for $n = 100$ and $\lambda = 1.0$

We can easily see from the table that under the given conditions, both the distributions give similar results. Thus it is proved that Poisson's distribution is a special case of binomial distribution.

Table 2.3: Comparison of Poisson and Binomial distribution

x	0	1	2	3	5
Binomial	.366	.3697	.1849	.0610	.0029
Poisson	.368	.3679	.1839	.0613	.0031

2.6 CONTINUOUS RANDOM VARIABLE

Suppose that you have purchased stock in *Colossal Conglomerate, Inc.,* and each day you note the closing price of the stock. The result of each day is a real number X (the closing price of the stock) in the unbounded interval $[0, +\infty)$. Or, suppose that you note time for several people running a 50-meter dash. The result for each runner is a real number X, the race time in seconds. In both cases, the value of X is somewhat random. Moreover, X can take on essentially any real value in some interval, rather than, say, just integer values. For this reason we refer to X as a **continuous random variable**. Here is the official definition.

A **random variable** is a function X that assigns to each possible outcome in an experiment a real number. If X may assume any value in some given interval (the interval may be bounded or unbounded), it is called a **continuous** random variable. If it can assume only a number of separated values, it is called a **discrete** random variable.

For instance, if X is the result of rolling a dice (and observing the uppermost face), then X is a discrete random variable with possible values 1, 2, 3, 4, 5 and 6. On the other hand, if X is a random choice of a real number in the interval $\{1, 6\}$, then it is a continuous random variable.

Example 2.6: The following table shows the distribution of UP residents (16 years old and over) attending college in 1980 according to age.

Age	15–19	20–24	25–29	30–34	35
Number in 1980 (thousands)	2,678	4,786	1,928	1,201	1,763

Draw the probability distribution histogram for X = the age of a randomly chosen college student.

Solution:

Summing the entries in the bottom row, we see that the total number of students in 1980 was 12.356 million. We can therefore convert all the data in the table to probabilities dividing by this total.

Age	$15 \le X < 20$	$20 \le X < 25$	$25 \le X < 30$	$30 \le X < 35$	≥ 35
Probability	0.22	0.39	0.16	0.10	0.13

The table tells us that, for instance,

$$P(15 \le X < 20) = 0.22$$

and

$$P(X \ge 35) = 0.13.$$

The probability distribution histogram is the bar graph we get from these data:

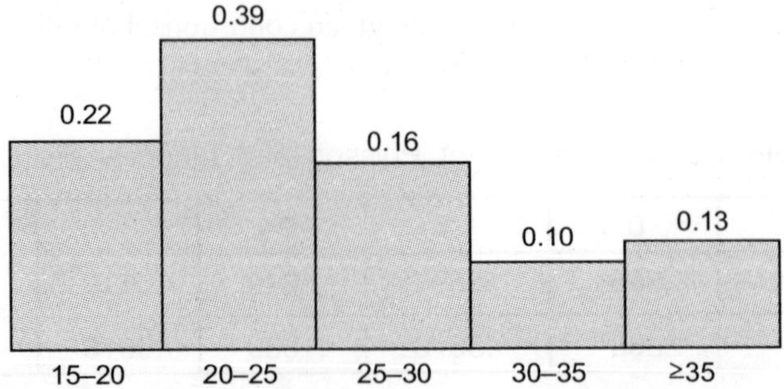

We can define continuous random variable in another way too. A random variable X and the corresponding distribution of random variable is to be continuous if the corresponding cumulative distribution function $F(x) = P(X \leq x)$ can be represented by an integral form.

$$F(x) = \int_{-\infty}^{x} f(x)dx \qquad ...(2.18)$$

where the integrand $f(x)$ is continuous everywhere except some of the finite values of x. The integrand $f(x)$ is called the probability density function or density of the distribution. In equation (2.18) we see that $F'(x) = f(x)$ for every x at which $f(x)$ is continuous, where $F'(x)$ is the derivative of $F(x)$ with respect to x. In this sense the density is the derivative of the cumulative distribution function $F(x)$. Since sum of all the probabilities in the sample space is equal to unity, from equation (2.18) one gets,

$$\int_{-\infty}^{+\infty} f(x)dx = 1 \qquad ...(2.19)$$

Since for any a and $b > a$, we have

$$P(a < x \leq b) = F(b) - F(a) = \int_{a}^{b} f(x)dx \qquad ...(2.20)$$

Thus probability $P(a < X \leq b)$ is the area under the curve with function $f(x)$ and between the lines $x = a$ and $x = b$ (see Fig. 2.7).

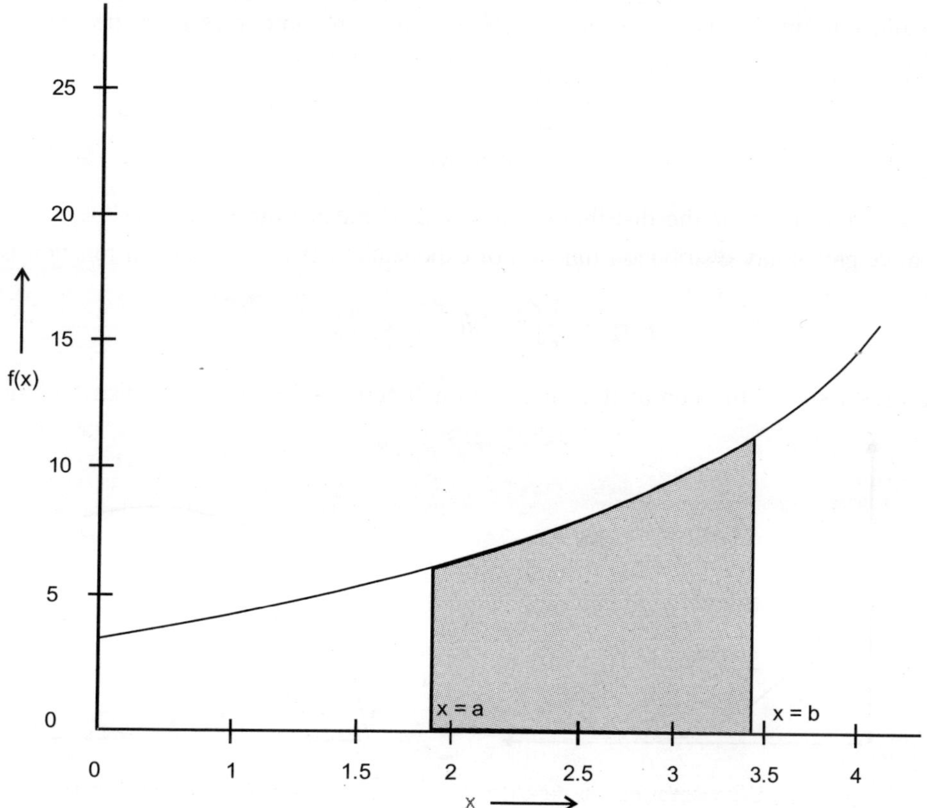

Fig. 2.7: *Probability function $P(a < x \leq b)$, is area under the curve f(x) and lines x = a and x = b.*

Obviously for any fixed a and b ($> a$) the probabilities corresponding to the intervals $a < X \leq b$, $a \leq X < b$, $a \leq X \leq b$, $a < X < b$ are all same. This is different from a situation in the case of a discrete distribution. Since the probabilities are non-negative and (2.18) holds for every interval, we must have $f(x) > 0$ for all x.

2.7 Exponential Distribution

Theory of queuing is an important subject in the field of Operational Research. Queues are generally found in banks, airports and even on roads. In such problems a customer has to stand in queue until he is served and allowed to leave. In such problems, probability that the next customer does not arrive during the interval t given that the previous customer arrived at time $t = 0$, is given by exponential density function. Seeing the importance of this density distribution function, it will be appropriate to discuss it here. A random variable X is said to be exponentially distributed with parameter $\lambda > 0$ if its probability distribution function is given by

$$f(x) = \begin{cases} \dfrac{e^{-x/\lambda}}{\lambda}, t \geq 0 \\ 0, \quad \text{elsewhere} \end{cases} \quad \text{...(2.21)}$$

The graphical representation of density function is shown in Fig. 2.8*a*. The mean and variance of exponential distribution has given by

$$E(X) = \mu = \lambda \text{ and } \mathrm{Var}(X) = \lambda^2 \quad \text{...(2.22)}$$

If variable x is time t, then exponential density function sometimes is written as,

$$f(x) = \begin{cases} \dfrac{1}{\tau} e^{-t/\tau}, t \geq 0 \\ 0, \text{ elsewhere} \end{cases} \quad \text{...(2.23a)}$$

where $\tau = \lambda$ is the mean of the distribution, also called mean time of arrival.

Cumulative probability distribution function of exponential density distribution function is given

$$F(t) = \frac{1}{\tau}\int_0^t e^{-t/\tau} dt = 1 - e^{-t/\tau} \quad \text{...(2.23b)}$$

Figure (2.8*b*) shows the cumulative distribution function of exponential distribution.

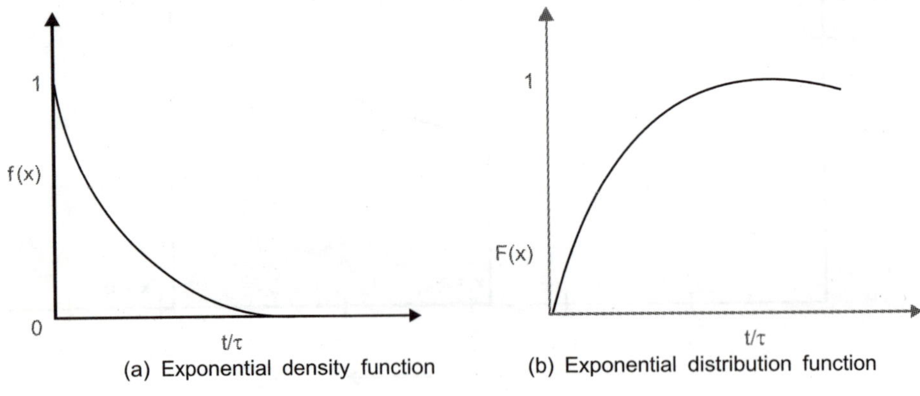

(a) Exponential density function (b) Exponential distribution function

Fig. 2.8: *Exponential distribution function.*

The exponential distribution has been used to model inter arrival times when arrivals are completely random and to model service times which are highly variable in the theory of queuing. Queuing theory will be discussed in chapter seven. This distribution is also used to model the life time of a component that fails catastrophically, such as light bulb. In this case λ is the failure rate.

2.7.1 Gamma Distribution Function

Exponential distribution function is a special case of Gamma distribution function for $\alpha = 1$, and $\beta = \lambda$, Gamma distribution function can be defined as,

$$f(x) = \frac{x^{\alpha-1}e^{-x/\beta}}{\Gamma(\alpha)\beta^{\alpha}} \begin{cases} x \geq 0 \\ \alpha, \beta \geq 0 \end{cases} \qquad ...(2.24)$$

where α, β are positive parameters and

$$\Gamma(\alpha) = \int_0^{\infty} x^{\alpha-1}e^{-x}dx \text{ for all } a > 0 \qquad ...(2.24a)$$

If α is an integer, then repeated integration of (2.24*a*), gives

$$\Gamma(\alpha) = (\alpha - 1)\,!$$

In this case (α an integer), Gamma distribution is called Erlang distribution. Below we determine the mean and variance of Gamma distribution. Mean μ is given by

$$\mu = E(X) = \int_0^{\infty} x\left[\frac{x^{\alpha-1}e^{-x/\beta}}{\beta^{\alpha}\Gamma(\alpha)}\right]dx = \int_0^{\infty}\left[\frac{x^{\alpha}e^{-x/\beta}}{\beta^{\alpha}\Gamma(\alpha)}\right]dx$$

Substituting $t = x/\beta$, we have

$$\mu = \frac{\beta^{\alpha}\beta}{\beta^{\alpha}\Gamma(\alpha)}\int_0^{\infty} t^{\alpha}e^{-t}dt = \frac{\beta}{\Gamma(\alpha)}\Gamma(\alpha+1) = \alpha\beta \qquad ...(2.24b)$$

$$E(X^2) = \int_0^{\infty} x^2\left[\frac{x^{\alpha-1}e^{-x/\beta}}{\beta^{\alpha}\Gamma(\alpha)}\right] = \int_0^{\infty}\left[\frac{x^{\alpha+1}e^{-x/\beta}}{\beta^{\alpha}\Gamma(\alpha)}\right]$$

$$\frac{\beta^2}{\Gamma(\alpha)}\Gamma(\alpha+2) = \frac{\beta^2}{\Gamma(\alpha)}\Gamma(\alpha+2) = \beta^2(\alpha+1)\alpha$$

Now since $\Gamma(\alpha+2) = (\alpha+1)\Gamma(\alpha+1) = (\alpha+1)\alpha\Gamma(\alpha)$

Therefore

$$\sigma^2 = E(X^2) - \mu^2 = \beta^2(\alpha+1)\alpha - (\alpha\beta)^2 = \alpha\beta^2 \qquad ...(2.24c)$$

When $\alpha = 1$ and $\beta = \lambda$, we get mean and variance for exponential distribution.

Example 2.7: Let the life of an electric lamp, in thousands of hours, is exponentially distributed with mean failure rate $\lambda = 3$. This means, there is one failure in 3000 hours on the average. The probability that lamp will last longer than its mean life of 3000 hours is given by $P(X > 3) = 1 - P(X \leq 3) = 1 - F(3)$. Equation (2.22*b*) is used to compute $F(3)$, thus

$$P(X > 3) = 1 - (1 - e^{-3/3}) = e^{-1} = 0.368$$

This result is independent of λ (since $x = \lambda$). Thus we observe that probability that bulb will last more than its mean value is 0.368, for any value of λ.

Probability that electric bulb will last between 2000 and 3000 hours is,

$$P(2 \le X \le 3) = F(3) - F(2)$$
$$= (1 - e^{-3/3}) - (1 - e^{-2/3})$$
$$= -0.368 + 0.513 = 0.145$$

Let us now discuss one of the most important property of exponential density distribution i.e., it is "memory less". That means, the probability that a component lives at least for $t + s$ hours, given that it has already lived for s hours, is same as the component at least lives for t hours, if life of component is exponentially distributed. Which means that for all $s \ge 0$ and $t \ge 0$,

$$P(X > s + t / X > s) = P(X > t) \qquad ...(2.25)$$

In equation (2.25), X represents the life of a component (a bulb say) and assume that X is exponentially distributed. Left hand side of the equation (2.25) states that the probability that the component lives for at least $s + t$ hours, given that it has already lived for s hours, and right hand side states that probability that it lives for at least t hours. If the component is alive at time s (if $X > s$) then the distribution of remaining amount of time that it survives, namely $X - s$, is the same as the original distribution of a new component. That is, the component does not "remember" that it has already been in use for a time s. A used component is as good as new.

That equation (2.25) holds is shown by examining the conditional probability

$$P(X > s + t / X > s) = \frac{P(X > t + s)}{P(X > s)} \qquad ...(2.26)$$

From equation (2.22*b*), we know that $F(t)$ is the probability that X is less than or equal to t. Thus probability that component lives at least for t hours is equal to $1 - F(t)$.

Substituting $1 - F(t) = e^{-t/\lambda}$ in the right hand side of (2.26), we get

$$P(X > s + t / X > s) = \frac{e^{-(s+t)/\lambda}}{e^{-s/\lambda}} = e^{-t/\lambda} \qquad ...(2.26a)$$

which is equal to $P(X < t)$.

Example 2.8: Find the probability that industrial lamp in example 2.7 will last for another 1000 hours, given that it is operating after 2500 hours.

Solution: We use equations (2.22) and (2.22*a*) and get,

$$P(X > 3.5 / X > 2.5) = P(X > 1) = e^{-1/3} = 0.717$$

This logically is not correct as time increases, probability of its functioning goes on increasing. For example probability of surviving for 6000 hours is nothing but $e^{-2.0} = 0.135134$.

2.7.2 Erlang Density Function

It has been mentioned while discussing Gamma density function that when α is an integer it becomes Erlang density function. A. K. Erlang was a Danish engineer responsible for the development of queuing theory. Let $\alpha = k$, and $\beta = \frac{\lambda}{k}$ in equation (2.24), where k is an integer. Equation (2.24) becomes

$$f(x) = (k/\lambda)^k \frac{x^{k-1} e^{-kx/\lambda}}{(k-1)!} \quad \begin{cases} x \ge 0 \\ k, \lambda \ge 0 \end{cases} \qquad ...(2.27)$$

For $k = 1$, $f(x)$ becomes,

$$f(x) = \frac{1}{\lambda} e^{-x/\lambda}, \; x \geq 0, \; \lambda \geq 0$$

which is nothing but exponential distribution.

2.8 Mean and Variance of Continuous Distribution

In the section 2.3, we explained what is expected value and variance of discrete random variable X. In this section we will discuss these functions for continuous random variable. The mean value or the mean of a distribution is denoted by μ and is defined by

$$(a)\ \mu = \Sigma\, z_j\, f(z_j) \text{ – discrete distribution} \qquad ...(2.27a)$$

$$(b)\ \mu = \int_{-\infty}^{+\infty} x f(x) dx \text{ – continuous distribution} \qquad ...(2.27b)$$

In (2.27*a*) $f(x_j)$ is the probability function of random variable X. The mean is also known as **mathematical expectation** of X and is denoted by $E(X)$. In case of continuous random parameter mean is nothing but integral over all the values of X.

A distribution is said to be symmetric with respect to a number c if for every real x, if

$$f(c + x) = f(c - x).$$

The variance of distribution is denoted by σ^2 and is defined by second moments as follows:

$$(a)\ \sigma^2 = \Sigma j\, (x_j - \mu)^2 f(x_j) \quad \text{(discrete distribution)} \qquad ...(2.28)$$

$$(b)\ \sigma^2 = \int_{-\infty}^{+\infty} (x-\mu)^2 f(x_j) dx \quad \text{(continuous distribution)}$$

The positive square root of the variance is called the standard deviation and is denoted by σ. Roughly speaking the variance is a measure of the spread or dispersion of the values which the corresponding random variable X can assume. Standard deviation can also be used to determine the dispersion of experimental data from the calculated data. It is nothing but square root of sum of square of experimental (observed) and actual values. This will be more clear in the following example.

Example 2.9: In Table 2.4 values of variable Y corresponding to values of X obtained in an experiment are shown. If in this data, we fit an equation $Y = 2X + 3$, find the standard deviation of the data values from the straight-line.

Table 2.4: Data obtained in an experiment

y_i	5	4	6	7	8	9	10
x_i	1.1	0.55	1.56	2.01	2.51	3.03	3.56
y	5.2	4.1	6.12	7.02	8.02	9.06	10.12

Solution: In the table x_i, y_i are experimental values and y are those obtained from the fitted expression. Variance of y thus is

$$s = \sum \frac{(y - y_i)^2}{n}$$

thus

$$s = \frac{(0.2)^2 + (0.1)^2 + (0.12)^2 + (0.02)^2 + (0.06)^2 + (0.12)^2}{10}$$

$$\therefore \qquad \sigma = \sqrt{s} = \sqrt{\frac{.0828}{10}} = 0.0909$$

Thus standard deviation of experimental data from a fitted curve is 9.09%.

2.9 Normal Distribution

Distribution of students scores in an examination, fragments of a shell or impact points of a gun, all follow a distribution called normal distribution. Normal distribution is a distribution discovered by Carl Friedrich Gauss (1777–1855). This distribution is extensively used in the study of target damage due to weapons. In nature, number of events follow normal distribution.

The continuous distributions having the density

$$f(x) = \frac{1}{\sigma\sqrt{2\pi}} \exp\{-(x-\mu)^2/\sigma^2\}, \sigma>0 \qquad ...(2.29)$$

is called the normal distribution or Gaussian distribution. A random variable having this distribution is said to be normal or normally distributed variable. This function is like an inverted bell shape being symmetric about point $x = \mu$ and is shown in Fig. 2.9.

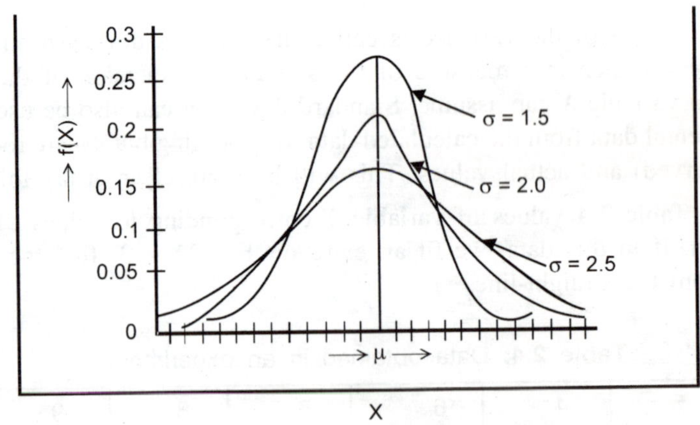

Fig. 2.9: *Shape of function f(x) for μ and different values of σ.*

This distribution is very important, because many random variable of practical interest are normal or approximately normal or can be transformed into normal random variables in a relatively simple fashion. In equation (2.29) μ is the mean and σ is the standard deviation of the distribution.

2.9.1 Properties of Normal Distribution Curve

(1) For $\mu > 0$ ($\mu < 0$) the curves have the same shape, but are shifted μ units right (to the left) of y-axis.

(2) The smaller σ^2 is, the higher is the peak at $x = \mu$.

2.9.2 Cumulative Density Distribution Function of Normal Distribution

The cumulative density distribution function of the normal distribution is

$$F(x) = \frac{1}{\sigma\sqrt{2\pi}} \int_{-\infty}^{x} \exp[-(x-\mu)^2/(2\sigma^2)]dx \qquad ...(2.30)$$

From equation (2.30), we observe

$$P(a < X \leq b) = F(b) - F(a) = \frac{1}{\sigma\sqrt{2\pi}} \int_{a}^{b} \exp[-(x-\mu)^2/(2\sigma^2)]dx \qquad ...(2.31)$$

This integral cannot be evaluated by elementary methods but can be represented in terms of the integral

$$\Phi(z) = \frac{1}{\sqrt{2\pi}} \int_{-\infty}^{z} e^{-u^2/2} du \qquad ...(2.32)$$

which is the distribution function with mean equal to zero and variance equal to one. This integral is obtained from integral (2.31) by substituting, $\frac{x-\mu}{\sigma} = u, \frac{du}{dx} = \frac{1}{\sigma}$, and limits for integration varying from $-\infty$ to $z = \frac{x-\mu}{\sigma}$.

Therefore equation (2.32) becomes

$$F(x) = \frac{1}{\sqrt{2\pi}} \int_{-\infty}^{\frac{(x-\mu)}{\sigma}} e^{-u^2/2} du \qquad ...(2.33)$$

The right hand side is same as that of equation (2.32) where

$$z = \frac{x-\mu}{\sigma}$$

$$F(x) = \Phi\left(\frac{x-\mu}{\sigma}\right)$$

Therefore combining eqs. (2.31) and (2.33) one gets

$$P(a < X \leq b) = F(b) - F(a) = \Phi\left(\frac{b-\mu}{\sigma}\right) - \Phi\left(\frac{a-\mu}{\sigma}\right) \qquad ...(2.34)$$

in particular, when

$$a = \mu - \sigma$$

and

$$b = \mu + \sigma$$

we get

$$P(\mu - \sigma < z \leq \mu + \sigma) = \Phi(1) - \Phi(-1)$$

Integral in equation (2.32) have been integrated numerically by using Simpson's method [4] and results are shown as follows

(*a*) $P(\mu - \sigma < x \leq \mu + \sigma) \approx 68\%$

(*b*) $P(\mu - 2\sigma < x \leq \mu + 2\sigma) \approx 95.5\%$

(*c*) $P(\mu - 3\sigma < x \leq \mu + 3\sigma) \approx 99.7\%$

A computer program for normal distribution function written in C++ language by Simpson's method is given below.

Program 2.2: Computer Program for Normal Distribution Function

```
#include <iostream.h>
#include <math.h>
#include <stdlib.h>
#include <conio.h>
#define func(x) exp(-x*x/2.0)
/* Function is defined in the beginning as global function. Any parameter
or function defined outside main is called global and can be called any where
in the program*/
int k,n;
float h,a,b,x;
double sum=0.0,suml=0.0,prev_result,acc=0.001,area,diff;
main( )
 {
cout<<'' This is a program to integrate the function exp(x^2/2)\n";
//Any thing in ""…"" means comment to be printed on the screen.
cout<<" \n\n Enter the lower limit of integral : ";
cin>> a;// Will read the value of a, the lower limit of integral.
Cout<<" \n Enter the upper limit of integral : ";
Cin>>b;
suml = func(a) + func(b);
prev_result = 0.0;
n = 2;
do
{
 h = (float)( b - a ) / (float)n;
 sum = sum;
 x = a;
 for(k = 1; k < n;k++)
     {
     x +=h;
     if((k%2) != 0)
     sum = 4.0*func(x);
     else
```

```
        sum += 2.0*func(x);
        }
        area = h*sum/3.0;
        diff = fabs(prev_result - area);
         prev_result = area;
         cout<<n<<'\t'<<h<<'\t'<<area<<'\t'<<diff<<'\t'<<acc;
    /*Here '/t' means give a space between the value of parameters n,
h,area,diff,acc.*/
         n +=2;
     }while (diff > acc);
     if (diff <= acc )
        {
         cout<<"\n \n");
         cout<<" integral of the function e(X^2) from a to b is \n";
         cout<<area;
         getch();
        }
        return;
    }
```

In the following paragraph we demonstrate an experiment on computer which demonstrates the normal distribution.

2.9.3 An Experiment for the Demonstration of Normal Distribution Function

To demonstrate the normal distribution function we conduct an experiment on computer. Let us assume that a soldier is firing at the centre of a target with his gun. Let the standard deviation of bullet landing on the target by the gun is σ. Draw circles of radii σ, 2σ and 3σ respectively on the target with centre at the *aim point*. Now if the soldier fires *n* (*n* being large) shots at the aim point then 0.46% (.68×.68%) of the shots will fall in the innermost circle and 91.2% (0.955 × 0.955%) will fall in the circle with radius 2σ and almost all the shots will fall in the circle of radius 3σ. On the other hand if 46% of the total shots fall within a circle whose radius is σ then σ is the standard deviation of the weapon. The same experiment now we conduct on the computer.

We know that distribution of shots follows normal distribution. On the screen of the computer we take a point and draw three circles of radii σ, 2σ, and 3σ taking this point as a centre. Then we generate two normal random numbers using two different seeds (For generation of random numbers see chapter five). These two numbers can be converted to (x, y) co-ordinates of a typical shot. Thus a point (x, y) is plotted on the screen of the computer. This process is repeated *n* number of times and *n* points are plotted. A counter counts the points falling in individual circle. Thus we get the above scenario (Fig. 2.10).

Hence we may expect that a large number of observed values of a normal random variable *X* will be distributed as follows:

(*a*) About 46% of the values will lie between and $\mu - \sigma$ and $\mu + \sigma$ and

(*b*) About 91% of the values will lie between $\mu - 2\sigma$ and $\mu + 2\sigma$

(*c*) About 99.1% of the values will lie between $\mu - 3\sigma$ and $\mu + 3\sigma$

In the simulation example numbers obtained in different circles are slightly different from theoretical values. This is because the total number of trials are not large (n = 2000). If n is increased, closer results will be obtained.

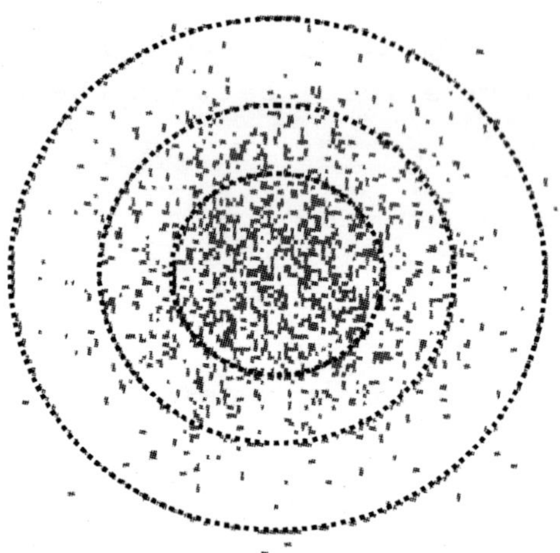

Fig. 2.10: *A computer output of normal distribution.*

Total number of hits = 2000

Hits in different circles : σ = 0.40, 2σ = 0.862, 3σ = 0.9890, CEP = 0.5055

2.9.4 Example of Dispersion Patterns

Consider a group of 10 rounds fired from a rifle at a vertical target. Let the co-ordinates of 10 impact point be

(1, 2), (–3, –1), (1, –1), (2, 4), (3, 0), (4, 3), (–1, 3), (2, –2), (–2, +1) and (3,1)

One can see that mean values of x and y of these points are

$$x_{\text{mean}} = \frac{1}{n}\sum x = \bar{x} = 1$$

$$y_{\text{mean}} = \frac{1}{n}\sum x = \bar{y} = 1$$

Therefore **Mean Point of Impact** (MPI) is (1,1) at distance of $\sqrt{2}$ from the origin. Take another group of 10 rounds, the MPI will be different from (1,1). Distance of MPI from the aim point is called the **aiming error** of the weapon. This error can be due to various reason. One of the reasons can be the error in the launching angle of the gun, or some other mechanical error in the weapon.

Thus the pattern of shots would vary from group to group in a random manner. The dispersion and MPI during the firing of a group of round will affect the probability of hitting a target, and the probability of damage which is very important in evaluating the effectiveness of a weapon.

2.9.5 Estimation of Dispersion

Sample variance in x-direction is the ratio of the sum of squares of the deviation of the x-co-ordinates from their mean to the number of impact points. Thus the sample variance in x-direction is given by

$$s_x^2 = \frac{1}{n}\sum_{i=1}^{n}(x_i - \bar{x})^2 \qquad ...(2.35)$$

Standard deviation S_x is the square root of the variance

Thus

$$S_x = \sqrt{\frac{1}{n}\sum_{i=1}^{n}(x_i - \bar{x})^2}$$

The standard deviation in this case is not the true standard deviation. It always has some error and is different from true standard deviation, which will be denoted by σ_x. We call this as biased standard deviation. But if the sample size is very large then this error reduces to zero. Thus true standard deviation is defined as,

$$s_x = \sqrt{\operatorname{Lim}_{n \to \infty} \frac{1}{n}\sum_{i=1}^{n}(x_i - \bar{x})^2}$$

If we reconsider the example of section 2.9.4,

$$\bar{x} = 1$$

$$S_x = \left[\frac{1}{10}\{0^2 + (-4)^2 + 0^2 + 1^2 + 2^2 + 3^2 + (-2)^2 + 1^2 + (-3)^2 + 2^2\}\right]^{1/2}$$

$$= 2.19$$

Similarly the sample variation in y-direction is

$$S_y = 1.897$$

and total sample variation would be

$$s = \sqrt{s_x^2 + s_y^2} = 2.898$$

If we consider a very large number of data points, then standard deviations S_x and S_y would approach to their true values σ_x and σ_x respectively.

It is known that sample variance is generally computed as [72]

$$S_x = \sqrt{\frac{1}{n-1}\sum_{i=1}^{n}(x_i - \bar{x})^2} \qquad ...(2.36)$$

which is based on $(n-1)$ degrees of freedom—one degree of freedom being used in the calculation of the sample mean $\bar{x}$ as an estimation of the population mean. This is for large values of n and is equal to σ_x^2. In fact both the sample variances converge to same value when n is large (see Appendix 4.2 : Sampling distribution of means).

The sample means and standard deviation describe only the location of the centre of data points and their dispersion. These two parameters do not however describe the characteristics of the overall distributions of data points. In case data points are hits on a target due to a weapon, we have to make some assumptions of reasonable practical value, in estimating probabilities of hitting. It is widely assumed and also verified sufficiently that the distribution of rounds is approximately normal or Gaussian in character. Thus, the probability density function of rounds in x-direction may be described by

$$f(x) = \frac{1}{\sqrt{2\pi\sigma_x^2}} \exp\left\{-\frac{(x-\bar{x})}{2\sigma_x^2}\right\}^2 \quad ...(2.37)$$

and that of the dispersion in y-direction by

$$f(y) = \frac{1}{\sqrt{2\pi\sigma_y^2}} \exp\left\{\frac{(y-\bar{y})}{2\sigma_y^2}\right\}^2 \quad ...(2.38)$$

where $\bar{x}, \bar{y}$ is true value of co-ordinates of mean point of impact.

The functions $f(x)$ and $f(y)$ are the symmetric bell-shaped distributions, depending on only two parameters—the mean and the standard deviation. These functions represent univariate or one directional distributions, whereas the dispersion we observe in firing is of bi-variate character.

If x and y are independent in the statistical sense and origin is the mean i.e., $\bar{x} = 0, \bar{y} = 0$ the appropriate bi-variate normal density function would be

$$f(x, y) = \frac{1}{2\pi\sigma_x\sigma_y} \exp\left\{-\frac{x^2}{2\sigma_x^2} - \frac{y^2}{2\sigma_y^2}\right\} \quad ...(2.38a)$$

If $\sigma_x \neq \sigma_y$, we call this as non-circular and if $\sigma_x = \sigma_y$ then it is circular bi-variate normal distribution and distribution is given by,

$$f(x, y) = \frac{1}{2\pi\sigma^2} \exp\left\{-\frac{x^2 + y^2}{2\sigma^2}\right\} \quad ...(2.38b)$$

which shows that (x, y) is the point normally distributed around origin.

2.10 Circular Probable Error (CEP) and the Probable Error (PE)

In this section we discuss various errors in case of weapon delivery which will be of interest to defence scientists. A measure of dispersion generally used to describe weapon delivery accuracy is the Circular Error Probable (CEP). Parameter CEP is the basic parameter for determining the error in weapon performance and is defined as the radius of the circle about the true mean point with respect to aim point; which includes 50% of the hits, considering a very large number of rounds fired onto the target area under stable firing conditions.

Given the circular normal density function (equation (2.38b))

$$f(x, y) = \frac{1}{2\pi\sigma^2} \exp[-(x^2 + y^2)/2\sigma^2] \quad ...(2.39)$$

we integrate this function over a circular target with the centre at the origin and equate the result to one-half

$$\iint_{x^2+y^2 \leq R_{0.5}^2} f(x, y)\,dx\,dy = 0.5 \quad ...(2.40)$$

where radius $R_{0.50}$ is radius of circle so that 50% of the hits fall inside it.

By making the transformations of variables

$$x = r\cos\theta,\ y = r\sin\theta,\ 0 \le \theta \le 2\pi$$

in the integral (2.40) one gets,

$$= \frac{1}{2\pi\sigma^2}\int_0^{R_{0.5}}\int_0^{2\pi} d\theta \exp[-r^2/2\sigma^2]\, r\, dr \qquad \text{...(2.41)}$$

$$= 1-\exp\left[-R_{0.5}^2/2\sigma^2\right] = 0.5$$

or

$$\exp\left[-R_{0.5}^2/(2\sigma^2)\right] = 0.5$$

Therefore $\qquad R_{0.5} = \text{CEP} = 1.1774\sigma$

or $\qquad \sigma = \text{CEP}/1.1774$

Hence σ can be defined in terms of CEP. This relation is true only for circular distribution. Integration of equation (2.41) is very difficult and so far no one has been able to solve this problem. If in the example of section 2.8.2, we draw a circle of radius CEP, almost 50% of the shots will fall within it.

2.10.1 Range and Deflection Probable Errors

If all the hits are projected onto a straight line in place of a point, the interval about both sides of the line which includes 50% of the shots is called the **Range Error Probable (REP)**. This error is equal to 0.6745 σ. The REP is a one dimensional or univariate measure of dispersion and is used commonly for range precision in firing. Similarly deflection probable error is the error in a direction transverse to range and is called **Deflection Error Probable (DEP)**. The value of DEP is also equal to 0.6745σ.

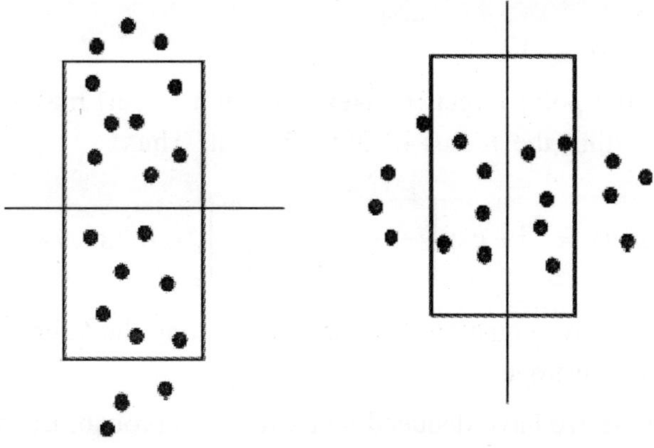

(a) Range error probable (b) Deflection error probable

Fig. 2.11: *Range error probable and deflection error probable.*

2.10.2 Probability of Hitting a Circular Target

In this section, we will discuss the probability of hitting a target, due to a weapon attack. Target can be an aerial or ground based. Various cases which effect the hit will be discussed.

Single Shot Hit Probability (SSHP) on a circular target of radius R with centre at the origin (this means there is no aiming error thus aim point coincides with the centre) will be considered. Circular normal distribution function is given by (equation (2.38*b*))

$$f(x, y) = \frac{1}{2\pi\sigma^2} \exp\left[-\frac{x^2 + y^2}{2\sigma^2}\right] \qquad ...(2.42)$$

Here σ is the standard deviation of weapon. Then the chance of a round hitting the circular target is simply (see section 2.10, equation (2.41))

$$p(h) = P(hit) = 1 - \exp\left[-R^2/2\sigma^2\right] \qquad ...(2.43)$$

Equation (2.43) only tells whether centre of lethal area of a particular weapon falls on a target whose radius is R or not and if it falls what is the probability. (This probability is also the chance that chi-square with two degrees of freedom is $\chi^2(2)$ and it does not exceed R^2/σ^2). To understand relation (2.43), let us consider the following example.

Example 2.10: Fire from an enemy bunker is holding up the advance of friendly troops. If an artillery with a warhead damage radius of 30m and delivery CEP of 20m is used what is the chance of destroying the bunker?

Here it is essential to tell that standard deviation σ is function of distance of aim point from the launch point of the weapon (Range) and is written as $\sigma = r\theta$, where θ is the dispersion of weapon in radians and r is the range. But if range is known beforehand, CEP can be given in terms of distance in place of angle.

Solution:

$$CEP = 20 = 1.1774\sigma$$

Therefore $\sigma = 17,$

The chance of killing the point target (bunker is a point target) may be found from the chance of a round falling on or within the radius of 30m from it. Thus

$$p(h) = 1 - \exp\left[-\frac{30^2}{2 \times 17 \times 17}\right] = 0.789774$$

It can be seen from the above equation that chance of hitting the target is 0.79 i.e., there are 79% chances that bunker will be destroyed.

Note that in this example we have deduced that with 79% probability, bunker will be destroyed. This logic does not seem to be correct, because target is a bunker, which is not a levelled target. Due to the depth of the bunker, it may not be possible to destroy it. In fact equation (2.34) only gives us that target is covered by the weapon.

EXERCISE

1. What is stochastic variable? How does it help in simulation? (PTU, 2004)
2. What is an exponential distribution? Explain with an example. (PTU, 2004)
3. Discuss in details, the discrete probability function. How it is different from continuous probability function? (PTU, 2003)
4. Suppose that a game is played with a single dice which is assumed to be fair. In this game player wins Rs.20.00 if a 2 turns up, Rs.40 if a 4 turns up and loses Rs.30 if a 6 turns up. For other outcomes, he neither loses nor wins. Find the expected sum of money to be won.

   (*Hint* : $E(X)$ = (0)(1/6) + (20)(1/6) + (0)(1/6) + (40)(1/6) + (0)(1/6) + (–30)(91/6))
5. The density function of a random variable X is given by

$$f(x) = \begin{cases} \frac{1}{2}x, & \text{for } 0 < x < 2 \\ 0, & \text{otherwise} \end{cases}$$

Find the expected value of X.

6. In a lottery there are 100 prizes of Rs.5, 20 prizes of Rs.20, and 5 prizes of Rs.100. Assuming that 10,000 tickets are to be sold, what is the fair price to pay for the ticket?
7. Find the expected value of a discrete random variable X whose probability function is given by

$$f(x) = \left(\frac{1}{2}\right)^x, \quad x = 1, 2, 3, ...$$

8. Generate three random variates from a normal distribution with mean 20 and standard deviation 5. Take n = 12 for each observation.

   Hint: A normal variate is given by (Central Limit Theorem)

$$y = \mu + \sigma\left(\sum_{i=1}^{n} r_i - \frac{n}{2}\right)$$

   Generate 12 uniform random numbers r_i and compute y.
9. Generate three random variates from an exponential distribution having mean value 8.

   Hint: A variate of exponential distribution is given as,

$$y = -\frac{1}{\mu}\ln(1-r) \text{ or } -\frac{1}{\mu}\ln(r)$$

   where $1/\mu$ is the mean of the distribution and r is the uniform random number.
10. A survey finds the following probability distribution for the age of a rented car.

Age	0–1	1–2	2–3	3–4	4–5	5–6	6–7
Probability	0.20	0.28	0.20	0.15	0.10	0.05	0.02

Plot the associated probability distribution histogram, and use it to evaluate (or estimate) the following:

(*a*) $P(0 \leq X \leq 4)$

(*b*) $P(X \geq 4)$

(*c*) $P(2 \leq X \leq 3.5)$

(*d*) $P(X = 4)$

11. Generate three random variates from an exponential distribution having mean value 8.

Hint: A variate of exponential distribution is given as,

$$y = -\frac{1}{\mu}\ln(1-r) \text{ or } -\frac{1}{\mu}\ln(r)$$

where $1/\mu$ is the mean of the distribution and r is the uniform random number.

12. There are 15 equally reliable semiautomatic machines in a manufacturing shop. Probability of breakdown per day is 0.15. Generate the number of break down for next seven days. Determine the mean and variance of the generated observations. Compare with the theoretical values of mean and variance.

13. The life time, in years, of a satellite placed in orbit is given by the following exponential distribution function,

$$f(x) = \begin{cases} 0.4e^{-0.4x}, & x \geq 0 \\ 0, \text{ otherwise} \end{cases}$$

(*a*) What is the probability of life of satellite being more than five years?

(*b*) What is the probability of life of satellite being between 3 and 6 years?

14. The distribution function for a random variable x is:

$$F(x) = \begin{cases} 3 - e^{-x}, & x \geq 0 \\ 0, & x < 0 \end{cases}$$

Find:

(*a*) Probability density function

(*b*) $P(x > z)$

(*c*) Probability $P(-3 < x \leq 4)$. (PTU, 2002)

15. Give expressions for Binomial, Poisson and Normal distributions. Under what conditions Binomial distribution is approximated by Poisson distribution. (PTU, 2002)

16. 5000 students participated in a certain test yielding a result that follows the normal distribution with mean of 65 points and standard deviation of 10 points.

(*a*) Find the probability of a certain student marking more than 75 points and less than 85 points inclusive.

(*b*) A student needs more than what point to be positioned within top 5% of the participants in this test?

(*c*) A student with more than what point can be positioned within top 100 students?

APPENDIX 2.1

PROOF OF *E* (*X*) AND *VAR* (*X*) IN CASE OF BINOMIAL DISTRIBUTION

PDE of Binomial distribution can be written as,

$$f(x,n,p) = C_x^n p^x q^{n-x}$$

where

$$\sum_{x=0}^{n} C_x^n p^x q^{n-x} = 1$$

Then

$$E(X) = \sum_{x=0}^{n} \sum x f(x,p,n)$$

$$= \sum_{x=0}^{n} x \frac{n!}{x!\ (n-x)!} p^x q^{n-x}$$

$$= np \sum_{x=1}^{n} \frac{(n-1)!}{(x-1)!\ (n-x)!} p^{x-1} q^{n-x}$$

since the value of term with $x = 0$ is zero. Let $s = x - 1$ in the above sum. Thus

$$E(X) = np \sum_{s=0}^{n-1} \frac{(n-1)!}{s!\ (n-1-s)!} p^s q^{n-1-s}$$

Now since

$$\sum_{s=0}^{n-1} \frac{(n-1)!}{s!(n-1-s)!} p^s q^{n-1-s} = (p+q)^{n-1} = 1$$

we get

$$E(X) = np$$

In order to compute variance Var(X) we first compute $E(X^2)$.

$$E(X^2) = \sum_{x=0}^{n} x^2 \cdot f(x,n,p)$$

$$= \sum_{x=0}^{n} x^2 \cdot \frac{n!}{x!(n-x)!} p^x q^{n-x}$$

$$= np \sum_{x=1}^{n} \frac{x \ (n-1)!}{(x-1)! \, (n-x)!} p^{x-1} q^{n-x}$$

Substituting $s = x - 1$ one gets

$$E(X^2) = np \sum_{s=0}^{n-1} (s+1) \frac{(n-1)!}{s! \ (n-1-s)!} p^s q^{n-1-s}$$

$$= np \sum_{s=0}^{n-1} (s+1) \ f(s, n-1, p)$$

But $$\sum_{s=0}^{n-1} (s+1) f(s, n-1, p) = \sum_{x=0}^{n-1} f(s, n-1, p)$$

$$= \sum_{x=0}^{n-1} f(s, n-1, p)$$

$$= (n-1) \ p+1$$

$$= np + q$$

Thus $$E(X^2) = np \ (np+q)$$

and we get

$$\text{Var}(X) = E(X)^2 - (E(X))^2 = n^2 p^2 + npq - n^2 p^2 = npq$$

❑❑❑

3

AN AIRCRAFT SURVIVABILITY ANALYSIS

Aircraft survivability analysis is an important study in the aircraft industry. It means, an aircraft when under enemy attack is capable of surviving or not. This study is useful for estimating the damage caused to the aircraft, as well as deciding the basic parameters of the aircraft, while under design stage. In chapter two we have studied probability distribution of discrete random events. Also various laws of probability have also been studied. In this chapter we will apply these laws for making a model of aircraft survivability. In chapter two we explained how circular normal distribution can be applied to calculate single shot hit probability. In this model these concepts will be further elaborated. Due to the complex nature of the aerial targets, their damage analysis is an independent field and it is necessary to deal it in a different chapter. Under aerial targets, various airborne vehicles such as aircraft, helicopter, Remotely Piloted Vehicles (RPVs) etc. can be categorised. There is an important term called 'denial criteria' in damage assessment problems. Denial criteria means how a target will be damaged, partially, completely and will be unserviceable. It has been observed from war data that even if an aircraft is hit and had several bullet holes, has come back from the battle field. Therefore the criterion that aircraft or any other target under attack is completely destroyed is very important and requires a separate study. A denial criterion for the aircraft is quite different from the ground targets and requires a detailed study. Aircraft consists of hundreds of vital parts and kill (totally damaged) of either of them can lead to aircraft's malfunctioning, which can lead to its kill after sometime of operation. It is quite important to study each and every vital part of the airborne vehicle critically, the energy required, and type of mechanism needed for its kill, all of which matter in this study. In this chapter we will discuss the basic mathematics required for modelling the aircraft vulnerability. Since aircraft is a moving object, and it has various capabilities to avoid enemy attack, the study of its survivability against air defence system is called, aircraft combat survivability study. Aircraft combat survivability **is defined as the capability of an aircraft to avoid and/or withstand** a man made hostile environment. This study is of quite importance during the design stage of aircraft, as well as under its operational stage. In the present chapter a general concept of susceptibility of aircraft, its vital parts and their kill mechanism alongwith various kill criteria will be discussed. A dynamic model of aircraft vulnerability will be taken up in chapter six. In this chapter only a static model of aircraft survivability, using probabilistic concepts learnt so far will be used to build the model. Here it will be assumed that projection of aircraft as well as its vital parts on a plane has been provided as inputs. In chapter six, we will study with the help of mathematical models, how to obtain projection of aircraft on a given plane. An aircraft has hundreds of vital parts but for the sake of simplicity, only three vital parts are considered in this model. More elaborate model will be discussed in chapter six.

3.1 Susceptibility of an Aircraft

Inability of an aircraft to avoid the radar, ground based air defence guns, guided missiles, exploding warheads and other elements that make up the hostile environment can be measured by parameter, P_H, called **Single Shot Hit Probability** (SSHP), and is referred as *susceptibility* of the aircraft. Single shot hit probability has been discussed earlier in the second chapter in details. In that chapter SSHP was defined for simple stationary targets only. SSHP evaluation for the moving targets is quite complex and will be studied in chapter six. In this chapter SSHP will be computed assuming aircraft as stationary, in order to demonstrate methodology for the kill of various vital parts and their effect on the kill of aircraft, and main stress will be given to criteria of aircraft kill mechanism. In this chapter only aircraft vulnerability is dealt but the same theory can be applied to other airborne bodies too.

Susceptibility of an aircraft can be divided into three general categories (Ball, Robert E):

(*a*) scenario

(*b*) threat activity and

(*c*) aircraft

The scenario means the type of environment in which encounter takes place. Threat activity means the enemy's weapon system which is used against the aircraft. Weapon system includes the ground detection and tracking system. Type of aircraft, its size, its manoeuvrability capabilities and other factors also play significant role in the susceptibility analysis. The susceptibility of an aircraft in general is influenced by following factors.

(*a*) **Aircraft design:** This factor is one of the important factors required for the susceptibility analysis, and all designers of new aircraft conduct susceptibility analysis during design stage itself. While designing aircraft, first consideration to keep in mind is, how to hide it from the enemy's eye. Small size of an aircraft helps it to reduce its susceptibility from radar's detection. If the engine of the aircraft does not release much smoke, it is better for reducing the susceptibility. Since engine radiates the heat, which can be detected by IR detector, it is advisable to hide the engine behind a part which is not so vital and has no radiation. Another factor is, good manoeuvrability capability of the aircraft to avoid the ground air defence weapons. There are some rolling manoeuvres, which when an aircraft adopts, can even avoid missile hit.

Table 3.1: Essential element analysis (EEA) summary

	Events and elements	EE	Questions
1.	Blast and fragments strike the aircraft	Yes	How many fragments hit the aircraft and where do they hit?
2.	Missile warhead detonates within the range?	Yes	Can the onboard Electronic Counter Measure (ECM) suite inhibit the functioning of the proximity fuse?

Contd...

3.	Radar proximity fuse detects aircraft	Yes	Will chaff decoy the fuse?
4.	Missile propelled and guided to vicinity of aircraft.	Yes	Can the target aircraft outmanoeuvre the missile?
5.	Missile guidance system functions in flight.	Yes	Are i.r. (infra-red) flares effective decoys?
6.	Missile guidance system locked on to engine (infra-red) i.r radiation.	Yes	Is the engine's infra-red suppresser effective in preventing lock on?
7.	Target's engines within missiles field of view?	Yes	Are the engine hot parts shielded?
8.	Enemy fighter manoeuvres to put target into field of view and within maximum range.	Yes	Does the enemy fighter have a performance edge? Does the target A/C have an offensive capability against the enemy fighter?
9.	Target aircraft designated to enemy fighter and fighter launched.	Yes	Does the on-board or stand off ECM suite have a communications and jamming capability?
10.	Fighter available to launch against the target.	Yes	Are there any supporting forces to destroy the enemy fighter on the ground?
11.	Enemy 'C3' net function properly.	Yes	Does the stand-off ECM suite have a communications jamming capability?

(*b*) **Tactics:** Tactics is another parameter which depends on pilot's skill and helps in reducing the susceptibility. To avoid detection by the enemy air defence system, pilot can hide his aircraft behind the terrains. This technique is called terrain masking. Also an aircraft sortie consisting of escorts aircraft to suppress enemy air defence, also helps in reducing the susceptibility. It can carry various survivability equipment, such as electronic jamming devices to jam the enemy detection system or weapon to destroy it. Designing aircraft in such a way that it has minimum Radar Cross Section (RCS) is an art of the day. Radar cross section of a target will be discussed in section 3.3. Stealth aircraft developed in USA have minimum radar cross section and are difficult to be detected. These points are explained by the following illustration of an aircraft on a mission to drop paratroopers in enemy's territory.

Example: Consider a case of a transport aircraft attempting to deliver troops on a bright sunny day to a location near the FEBA (Forward Edge of Battle Area).

- It drops down into a valley to take advantage of terrain masking,
- A self propelled RADAR - directed Anti Aircraft Artillery (AAA) system detects the aircraft with the scanning radar through its optical tracker.
- Meanwhile RWR (Radar Warning Receiver) in aircraft detects scanning signals from AAA (Anti Aircraft Artillery) radar and alerts pilot for the type, location and status of the THREAT.
- Pilot ejects chaff, attempt to break the lock of tracking radar by manoeuvring and looks for hide out (terrain or vegetation).
- Observer on the AAA platform, watching the aircraft, redirects radar towards it.
- After short time, when a fire control solution is obtained ground system fires at aircraft.

The MODELING and quantification of individual events and elements in such an encounter is referred as a SUSCEPTIBILITY ASSESSMENT. Obviously, there are many diverse factors that influence susceptibility, many of which are difficult to model and to quantify. In order to determine which of the factors or events and elements are the most important and which ones are of lesser importance, an Essential Elements Analysis (EEA) should be conducted. To illustrate the EEA, consider an encounter between a friendly strike aircraft and an enemy interceptor carrying an air-to air, infrared homing missile. Details of the analysis is given in Table 3.1.

3.2 THREAT EVALUATION

Identification of enemy threat against the aircraft is also one of the important factor in the study of aircraft vulnerability. Threat element in general can be classified as terminal and non-terminal. The non-terminal threat are such, which do not create itself some damage to the aircraft, but they help the terminal threat to inflict damage. Under non-terminal threat come electronic and optical systems which are used for the support of terminal threat. Under this category fall surveillance, tracking and early warning radar, Electronic-Counter-Counter Measure (ECCM), fire or weapon control and communication units. These equipment are integral part of enemy's offensive and defensive weapon units. Terminal threat units are nothing but actual weapon such as missiles, anti aircraft guns etc., capable of inflicting damage to the aircraft.

3.3 SUSCEPTIBILITY ASSESSMENT (MODELING & MEASURES)

A susceptibility assessment is a modelling of the sequence of events and elements in the encounter between the aircraft and the THREAT until there are one or more hits on the aircraft body.

Events and elements identified in the EEA as per their importance should be included in the model. It is generally not easy to include all the events in the model. One of the parameter to be determined in susceptibility analysis is probability of detection P_d, which will be discussed briefly in the following paragraphs.

Aircraft Signature: The characteristics of the aircraft that are used by the threat elements for detection and tracking are called the aircraft signature. Some of the important aircraft signatures are,

– Radar Signature
– IR Signature ⎯ electro-magnetic
– Aural (Sound) Signature

Out of these signatures, radar and Infra-Red (IR) signatures are of electromagnetic type.

Let us understand how a target is detected by radar. Signal from ground radar strikes the aircraft, a part is absorbed as heat, a part may be reflected or scattered from various parts of the aircraft.

Total portion of the impinging signal eventually reflected in the direction of receiving radar is known as the aircraft radar signature σ. The size of the signature is referred as the aircraft's Radar Cross Section (RCS). In fact σ is a very complex parameter (Unit = 1m^2).

RCS of Aircraft : In the theory, the re-radiated or scattered field and thus RCS can be determined by solving Maxwell's equations with proper Boundary Conditions (BCs) for a given target. However, this can only be accomplished for most simple shapes. We derive below the radar equation and explain how radar detects a target.

Let us assume that a radar transmits a pulse with power P_t, towards a target which is at a distance R from it. Assuming that the radar transmitting antenna transmits power isotropically in all the directions, with spherical symmetry (area of the surface of enveloping sphere being $4\pi R^2$), then power flux per unit area at range R will be

$$\text{Power density} = \frac{P_t}{4\pi R^2}$$

If the radar antenna has a gain G, then the power density at the target will be multiplied by G. Now assume that target reflects back all the power intercepted by its effective area, and the reflected pattern is also isotropic. If the reflected area of the isotropic target is σ, which is called the **radar cross-section**, then the power reflected isotropically from the target is given by,

$$\text{Reflected echo signal} = \frac{P_t G}{4\pi R^2}\sigma$$

Radar cross-section σ has units of area and can be measured experimentally. Since the power is reflected isotropically, the reflected power density received by the antenna is

$$\text{Reflected power density} = \frac{P_t G}{4\pi R^2}\cdot\frac{\sigma}{4\pi R^2}$$

If the radar's receiving antenna has an effective area A, then the power received by the antenna is given by,

$$P_R = \frac{P_t G A}{4\pi R^2}\cdot\frac{\sigma}{4\pi R^2} \qquad \text{...(3.1)}$$

Also it is known that relation between gain G, and effective area A is [32]

$$A = \frac{G\lambda^2}{4\pi} \qquad \text{...(3.2)}$$

Substituting A in (3.1) from (3.2), one gets

$$P_R = \frac{P_t G^2\lambda^2\sigma}{(4\pi)^3 R^4} \qquad \text{...(3.3)}$$

In the derivation of radar's equation (3.3), it has been assumed that reflections from the target are isotropic, which is only an assumptions and not happens in actual situation. But still these equations can be used. We replace isotropic target for original target and change the area of isotropic target in such a manner that it gives same reflection echo, as original target gives. The area of the equivalent isotropic target calculated in this way is called the radar cross-section of the target, which off course is different from the actual cross-section of the target. Here in radar equation, noise factor has not been considered. When effect of signal to noise ratio is also taken into account, equation (3.3) becomes [32,71]

$$R_{\max} = \left(\frac{P_t G^2\lambda\sigma F^4}{(4\pi)^3 L_s \xi_{m/n} N L_a}\right)^{1/4} \qquad \text{...(3.3}a\text{)}$$

where

L_s = radar system losses ($L_s \geq 1$),
F = relative electrical field strength of echo at the receiving antenna,
N = noise of internal system or detection process,
L_a = signal and echo loss due to radar transmission in the atmosphere (≥ 1).

ξ_{min} = (S/N) = signal to noise ratio associated with a specified probability of detection.

For the derivation of this equation, reader is advised to refer some book on radar [71].

3.3.1 Aircraft Detection and Tracking

Air-defence systems utilise several procedures to detect, identify and track aircraft. These procedures are usually based upon a timewise progression of accuracy of aircraft location, right from warning by radar determining aircraft's location in azimuth, elevation, velocity and range as function of time. Below we discuss three major types of detection and tracking systems i.e., radar directed, the infrared and the visually directed systems.

Radar-directed detection systems

Air defence radar are of two categories i.e., detection or surveillance radars and weapon or fire control radar. Surveillance radar are usually pulse radar and are used for surveillance purpose.

Maximum Range: The maximum distance of a surveillance radar antenna at a height h_{ant} can see an aircraft at a altitude h_{ac} is limited by the radar horizontal range R_h in nautical miles is given by $R_h = 1.23(\sqrt{h_{ant}} + \sqrt{h_{ac}})\ (n{\cdot}m_i)$

The maximum range at which a radar operator can recognise an aircraft as a target is given by radar range equations (3.3a) [32,71].

3.3.2 Probability of Detection

During each scan of the target by the radar beam, a large number of pulses will be transmitted and echoes received. The receiver will either process these echoes individually, or several of the echoes will be summed up to improve the probability of detection.

Single look probability of detection $P_d(s)$, which is a function of S/N ratio is given by [71]

$$\bar{P}_d(s) = \int_0^{v} P(v)dv = \int v \exp\left\{\frac{-v^2 - 2\cdot\frac{S}{N}}{2}\right\} I_0\left[\sqrt{\frac{2S}{N}}\cdot v\right]dv \qquad ...(3.3b)$$

This equation when false alarm is taken into account reduce to,

$$\bar{P}_d(s) = 1 + e^{-S/N}\int_1^{P_n} I_0\left\{\sqrt{-4\frac{S}{N}\ln u}\right\}du$$

where

$$u = \exp\left\{\frac{-v^2}{2}\right\},$$

$$v = \sqrt{-2\ln u},$$

$$p_n = \exp\left\{\frac{-E_t^2}{2}\right\},$$

v = envelope of signal plus noise,

S/N = signal t noise power ratio,

I_0 = hyperbolic Bessel function of zero order,

$P(v)$ = probability density of v,

E_t = threshold limit of v.

For the s-th scan, there will be some probability of detection $P_d(s)$ based upon actual signal to noise ratio, which can be determined from the radar equations and cross-section, both of which can vary with each scan. Thus $P_d(s)$ given by equation (3.3b) is generally a function of time. If we denote this by P_d^t then probability that target has been detected after s scans is given by

$$\overline{P}_d(s) = 1 - \prod_{t=1}^{s}(1 - P_d^t)$$

3.3.3 Infra-red, Visual and Aural Detection

Apart from the detection by radar, aircraft can also be detected by its infra-red signature. Many aircraft are shot down after visual detection. Even aural detection some time is sufficient for shooting down the aircraft. Generally aircraft's sound can be heard about 30 seconds before it is actually visible. Smoke released by aircraft can be detected from a long distance and can be fatal for the aircraft. These can also be modelled as part of probability of detection.

3.4 Vulnerability Assessment

In broader sense, vulnerability of an aircraft can be defined as one minus survivability. The measure of vulnerability is the conditional probability, that the aircraft is killed given a hit on it and is denoted by a symbol $P_{k/h}$. Evaluation of parameter $P_{k/h}$ is quite complex and is the subject matter of this as well as chapter six. To determine kill to a vital part of an aircraft, first step is to find whether it is hit by the weapon or not. Weapon can be a Direct Attack (DA) shell or a fragment of shell fitted with proximity fuse. Once it is hit then various other factors, like penetration in the skin of that particular part and type of kill of the part, will arise. Whether that part is vital or not, if vital how performance of aircraft is effected by its kill is an important question. In order to achieve this, complete knowledge of structural data of aircraft and its vital parts is required.

By structural data we mean co-ordinates of all the points on aircraft, their thickness and material properties have to be given in the form of data tables. To achieve this, aircraft's structure is divided into finite number of triangles, co-ordinates of each triangle being known with reference to some fixed point of the aircraft. This point may be the nose of aircraft or its geometric centre. These co-ordinates arranged in a proper way are fed to computer to simulate a three dimensional view of the aircraft. Since aircraft is moving, these co-ordinates are moving co-ordinates. In order to determine single shot hit probability due to a ground based weapon, these co-ordinates are to be transferred in terms of co-ordinate system fixed with respect to the ground. Transformation of co-ordinates and projection of aircraft on a plane normal to the path of the trajectory of the weapon will be discussed in sixth chapter. In this chapter it will be assumed that aircraft projections on a given plane have been provided as input to the model.

Some of the important information, required for the vulnerability studies are:

1. **Categorisation of kill levels:** Kill of an aircraft can be expressed in different ways. One is to express the kill in terms of numbers. We can say that there is 70% probability that aircraft is killed, but this does not explain the exact nature of kill. Or it can also be expressed in terms of probability that an aircraft has a given level of kill. Kill level depends on the time taken for the repair of the aircraft to again put it into service. If aircraft is damaged beyond repair, we say that kill is catastrophic or KK-type of kill, In this type of kill an aircraft is disintegrated immediately after hit. Other types of kill levels are,

   K - kill: Damage caused, when aircraft is out of manned control in 20 seconds.

   A - kill: Damage caused to an aircraft that it falls out of manned control in five minutes, after being hit.

B - kill: Damage caused to an aircraft that it falls out of manned control in 30 minutes, after being hit.

These kills will be discussed in greater details in sixth chapter.

2. **Aircraft description:** Another factor is the assembly of the technical and functional description of the vital parts of the aircraft. That is, to list all the vital parts, their location, size, material function and criteria of their being killed. The functional description should define the functions provided by each component including redundancies. Vital part, sometime called critical component is defined as a component which, if either damaged or killed, would yield a defined or definable kill level of aircraft. For example, pilot is a critical component for KK-type of kill of an aircraft. If pilot is killed by a fragment or projectile, there is no hope of aircraft's survivability. Similarly engine in a single engine aircraft is a critical component for the A type kill of an aircraft. But for twin engine, it is not that much critical component. It is possible that both the engines are damaged by fragment hits. In that case even redundant engine become critical components. To find the kill of aircraft due to the damage of it's critical component a Critical Component Analysis (CCA), in tabular form is to be conducted. If a particular critical part is killed, what will be the effect on overall performance of the aircraft. This will involve listing of all the critical components and their roll in the performance of the aircraft. This process is called critical component analysis. Thus for critical component analysis, the first step is to identify the flight and mission essential functions that the aircraft must perform in order to continue to fly and to accomplish mission.

3.5 Vulnerability Due to Non-explosive Penetrator

To evaluate the vulnerability of an aircraft, which is coming toward a friendly target for attack, by ground based air defence system, involves study of number of parameters. In this as well as in sixth chapter, this subject will be discussed extensively. Apart from probability of detection $\overline{P}_d(s)$, first and foremost parameter in vulnerability study is the Single Shot Hit Probability (SSHP) of a target i.e., probability that a shot fired towards the target lands on it. SSHP has been studied in chapter two section 2.9.2, for stationary ground targets. For a target which is moving with high speed, the method given in this and second chapter will not work. This topic will be discussed in sixth chapter. In the present chapter a simplified method of evaluation of SSHP, assuming that target is stationary, will be given. SSHP of *i*-th component of an aircraft is assumed to be the ratio of projected areas of the component to that of aircraft. Also it is assumed that probability of kill of a component, subject to a hit on it is a given input. But in the sixth chapter, this parameter will be evaluated for different class of weapons.

Probability of hit on the *i*-th part of aircraft is defined as,

$$P_{h/H_i} = A_{P_i}/A_P \qquad ...(3.4)$$

Here P_{h/H_i} is probability of hit on the *i*-th part of aircraft, when bullet is aimed at the centre of it denoted by H, A_{P_i} is the projected area of *i*-th part on a given plane, and A_P is the projected area of the aircraft on the same plane. The typical plane is the plane normal to the path of bullet. Probability of hit here is defined as ratio of projected areas of typical part to that of aircraft. Other measure of vulnerability to impacting damage mechanism is defined as the aircraft vulnerable area A_v

Vulnerable area A_{v_i} of *i*-th critical component is defined as

$$A_{v_i} = A_{P_i} P_{k/h_i} \qquad ...(3.5)$$

where A_{p_i}, P_{k/h_i} are respectively the projected area of *i*-th component on a given plane and probability of kill of *i*-th component, subject to a hit on it. In the present model P_{k/h_i} will be assumed to be given input. Here after, all the parameters ending with capital subscripts denote values for aircraft and that ending with lower characters give values for component. Kill probability of *i*-th vital component given a random hit on the aircraft is given as,

$$P_{k/H_i} = P_{h/H_i} P_{k/h_i} \qquad ...(3.6)$$

where

P_{k/H_i} = probability of kill of *i*-th vital part, subject to a hit on aircraft,

P_{h/H_i} = probability of hit on the *i*-th vital part, subject to a hit on the aircraft,

P_{k/h_i} = probability of kill of *i*-th vital part, subject to a hit on it.

Equation(3.6) can be read as, probability of kill of *i*-th part of aircraft subject to a hit on aircraft is equal to probability of hit on the *i*-th part of aircraft subject to a hit on aircraft and probability of its kill subject to hit on it. Here projectile is aimed at the geometric centre of the aircraft. Probability of hit of *i*-th vital part is given for simplicity, by the ratio of the projected area of the vital part and aircraft respectively i.e.,

$$P_{h/H_i} = A_{p_i}/A_P \qquad ...(3.7)$$

where A_{p_i}, A_P are respectively are projected areas of vital part and aircraft, on the given plane. It is important to mention here that in order to find projected area of a component or of full aircraft, large number of transformations is needed, which have not been explained here. This algorithm works, when projected areas on a plane have been provided as part of data. In chapter six, a detailed algorithm for determining the projected areas on any plane is given. Since in actual dynamic conditions, aircraft, projectile and its fragments after explosion, all are moving with respect to each other, a series of projections are to be evaluated for determining a hit.

From equation (3.6) with the help of eqs. (3.5) and (3.7), one gets the probability of kill of *i*-th vital part, subject to a random hit on the aircraft as,

$$P_{k/H_i} = \frac{A_{p_i}}{A_P}\frac{A_{v_i}}{A_{p_i}} = \frac{A_{v_i}}{A_P} \qquad ...(3.8)$$

which is the ratio of vulnerable area of *i*-th vital part to projected area of whole aircraft.

It is to be noted in equation (3.5) that $A_{v_i} = A_{p_i} P_{k/h_i}$, which assumes that P_{k/h_i} is the given input. From equations (3.5) and (3.8) one gets

$$\frac{A_{p_i}}{A_P} = \frac{P_{k/H_i}}{P_{k/h_i}}$$

In the sixth chapter, we will show that this assumption is too vague, as given a hit on the vital part does not necessarily lead to a kill. Kill depends on the type of weapon and energy released by it on the aircraft surface. Some of the vital parts are hidden behind other non-vital parts. A projectile has to penetrate all these part and reach the vital part with sufficient critical energy required to kill it. Now we define an other term, that is probability of survival of aircraft. Probability of survival of aircraft $P_{S/H}$ is defined as one minus probability of its kill, thus

$$P_{S/H} = 1 - P_{K/H} \qquad ...(3.9)$$

Let us assume the aircraft has n vital parts. It is known that if one of the parts is killed, aircraft cannot fly. Thus for an aircraft to survive an enemy threat, all its critical components have to survive i.e.,

$$P_{S/H} = P_{s/H_1} \times P_{s/H_2} \times \; ... \; \times P_{s/H_n} = \prod_{i=1}^{n} P_{s/H_i} \qquad ...(3.10)$$

where

$P_{S/H}$ = probability of survival of aircraft subject to a random hit on it.

P_{S/H_i} = probability of survival of i-th vital component subject to a hit on the aircraft.

For the sake of simplicity, we assume that aircraft has only three critical (vital) components i.e., engine, pilot and fuel tank. Equation (3.10) with the help eq. (3.9) in this case reduces to,

$$P_{S/H} = (1-P_{k/H_1})(1-P_{k/H_2})(1-P_{k/H_3}) \qquad ...(3.11)$$

By multiplying terms on the right side of this equation one gets

$$P_{S/H} = 1 - \sum_{i=1}^{3} P_{k/H_i} + \sum_{i,j=1}^{3} P_{k/H_i} P_{k/H_i} - P_{k/H_1} P_{k/H_2} P_{k/H_3} \qquad ...(3.12)$$

Throughout this chapter, we will consider an aircraft with three vital components. Below we will illustrate these equations with numerical examples.

Case-1: To enhance the survivability of aircraft, generally more vital components are concealed behind less vital components to protect them. In this section we will assume that vital components are open to attack and then conceal them behind each other to see the effect of overlapping. During flight of aircraft vis a vis projectile, positions of various components keep on changing. Let us first consider a case when vital components of the aircraft are non-overlapping and non-redundant. By non-overlapping we mean, while projecting aircraft on a plane, projection of two components of aircraft are not overlapping. This condition is only a hypothetical condition as, if in one orientation two components are not overlapping, does not mean that they will not overlap in other orientation. Actually plane on which projection is taken is a plane, normal to the flight path of the projectile. This plane is called normal plane (N-plane). The condition of overlapping of parts varies from fragment to fragment. Now if components are non-overlapping, then more than one component cannot be killed by a single penetrator i.e., only one P_{k/H_i} will be non-zero, other two will be zero. Here $P_{k/H_i} = 0$ means, i-th part is not killed, subject to hit on the aircraft.

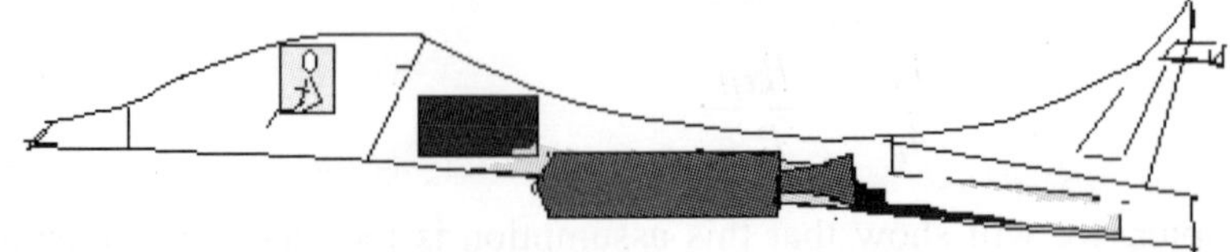

Fig. 3.1: *Aircraft with three non-redundant non-overlapping parts.*

Thus all the products of kill probability are zero, thus equation (3.12) reduce to,

$$P_{S/H} = 1 - \sum_{i=1}^{3} P_{k/H_i}$$

let us try to understand equation (3.12) by an example. This situation is explained in Table 3.2, where projected areas of aircraft and their vital components have been provided in meters, as a given input.

Table 3.2: Case of non-redundant and non-overlapping parts

Critical components	A_{p_i}, m²	P_{k/h_i}	A_{v_i}	P_{k/H_i}
Pilot	0.4	1.0	0.4	0.0133
Fuel (fire)	6.0	0.3	1.8	0.0600
Engine	5.0	0.6	3.0	0.1000
	A_P = 30.0 m²		A_v = 5.2	$P_{K/H}$ = 0.1733

In Table 3.2, we have used the relations (3.7) and (3.8), i.e.,

$$A_{vi} = A_{pi} \cdot P_{k/h_i}; \frac{A_{v_i}}{A_p} = P_{k/H_i} \qquad \text{...(3.12}a\text{)}$$

In this table, in third column P_{k/h_i}, have been assumed based on practical experience. For example, if pilot is killed, aircraft cannot fly and thus kill of pilot is taken as 100% kill of aircraft, that is $P_{k/h_i} = 1.0$. Same way other probabilities have been provided. A_P in the last row is equal to the total projected area of aircraft which is again a given input. First column of the table gives projected areas of three vital parts on a given plane, and column two gives their kill probabilities subject to a hit. Here it is to be noted that a hit does not always result in total kill. There are various other factors, such as total kinetic energy of the projectile. Column three and four are obtained by equation (3.12*a*). Total kill probability of aircraft in this case is sum of the kill of three vital parts and is 17.33% and total vulnerable area is 5.2 metre square.

3.5.1 Case of Multiple Failure Mode

In some aircraft, engine and fuel tank are close to each other. It is quite possible that by the rupture of fuel tank by the projectile hit, fuel may spill over the engine, which is quite hot and may catch fire. This type of kill is called multi-mode kill. Thus a hit on one component affects the vulnerability of other component. Let us see how this case is modelled?

Let probability of kill of aircraft by fire in fuel tank = 0.3 and probability of its kill due to ingestion of fuel on engine = 0.1. Since these two failures are not mutually exclusive i.e., if one event occurs, other is bound to occur, the aircraft will survive only if both survive i.e., $P_{S/H} = P_{S/H_f} \cdot P_{S/H_e}$. In such a case probability of kill of aircraft is given by,

$$P_{K/H} = 1 - P_{S/H_f} \cdot P_{S/H_e}$$

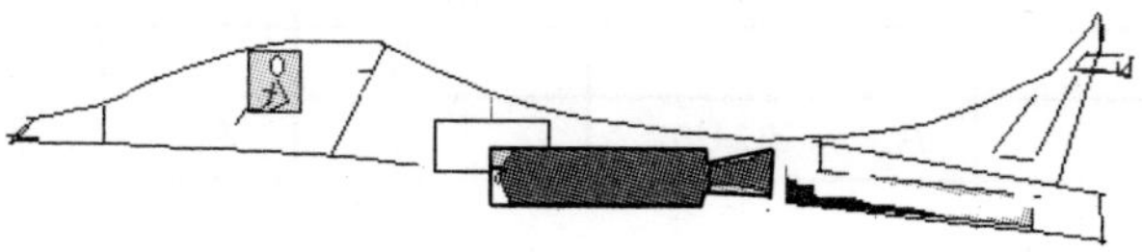

Fig. 3.2: *Case when two components overlap.*

Therefore, probability that aircraft's loss is due to fire or/and fuel ingestion is = 1 – (1 – 0.3) × (1 – 0.1) = 1 – 0.63 = 0.37. Thus probability of fuel tank has increased from 0.3 to 0.37 due to multiple failure modes (Table 3.3). Thus we see that in order to reduce the kill of the aircraft, engine and fuel tank should have sufficient separation. It is to be noted that this kill probability is not the simple sum of two probabilities. Reason for this is that due to single hit both the events may or may not occur. Therefore the probability of not occurring of both the events is first calculated and then probability of occurrence of at least one of the event is calculated.

Table 3.3: Case of non-redundant and non-overlapping parts (multi-mode failure)

Critical components	A_{P_i} m²	P_{k/h_i}	A_{v_i}	P_{k/H_i}
Pilot	0.4	1.0	0.4	0.0133
Fuel (fire)	6.0	0.37	2.22	0.0740
Engine	5.0	0.6	3.0	0.1000
	A_P = 30.0 m²		A_v = 5.2	$P_{k/H}$ = 0.1873

Thus in Table 3.3, kill probability due to fire in fuel tank increases from 0.3 to 0.37 and total kill probability of aircraft increases from 0.1733 to 0.1873 (Table 3.3). Thus here we see the benefit of simple modelling, which gives direct conclusion that engine and fuel tank in an aircraft should be at safe distance.

3.6 Case of Non-redundant Components with Overlap

The concept developed so far will further be extended in this section for more complex case. Now consider a case when two components are overlapping i.e., they are on the same shot line (line joining projectile centre and vital part's centre) and are killed by the same projectile, together. This is possible only if projectile has sufficient energy, so that after penetrating front part, it is able to penetrate part hiding behind it. Let overlapped area of both the components be denoted by A_{p/h_o}. If this overlapped area is killed then both the vital parts will be killed. Similarly, if this area has to survive, then both the vital parts should survive. Thus probability of kill of this area is = 0.72 (see Table 3.4). If A_{p/h_o} = 1.0m², then probability of kill of aircraft $P_{K/H}$ is given in the Table 3.4. It is seen from the Table 3.4 that now we have four vital parts in place of three. That is overlapped area of fuel tank and engine is taken as fourth vital part. Thus projected area of engine and fuel tank is reduced by one meter square.

Table 3.4: Case of non-redundant and overlapping components

Critical components	A_{P_i}	P_{k/h_i}	A_{v_i}	P_{k/H_i}
Pilot	0.4	1.0	0.4	0.0133
Fuel	6.0 = 1.0	0.3	1.5	0.050
Engine	5.0 = 1.0	0.6	2.4	0.0800
Overlapped area	1.0	0.72	0.72	0.024
	A_p = 30.0m²		A_v = 5.02	$P_{k/H}$ = 0.1673

$P_{k/H_o} = 1-(1-0.3)(1-0.6) = 0.72$ *(No Fire considered)*

From Table 3.4 it can be seen that overall probability of kill reduces, due to overlapping of two components. Due to overlapping of vital components it is quite possible that component ahead is partially pierced and component behind it is not damaged. This will further reduce vulnerability. It is a normal practice in modern aircraft to hide more vital components behind less vital components to reduce its vulnerability. In this section we have assumed that there was no fire in engine due to ingestion of fuel on the hot engine and thus engine did not received damage due to fire. In the next section, we will extend this model by assuming that engine catches fire due to ingestion of fuel.

3.6.1 Area with Overlap and Engine Fire

Since engine is overlapped, it can catch fire due to fuel ingestion as was shown in section 3.5.1. Let probability of kill of overlapped part of engine is 0.9. As earlier probability of kill of overlapped part is,

$$P_{k/h_o} = 1 - (1 - 0.3)(1 - 0.9) = 0.93$$

Results of this case are shown in the Table 3.5.

Table 3.5: Area with overlap and engine fire

Critical components	A_{P_i}	P_{k/h_i}	A_{V_i}	P_{k/h_i}
Pilot	0.4	1.0	4	0.0133
Fuel	6.0 = 1.0	0.3	15	0.050
Engine	5.0 = 1.0	0.6	24	0.0800
Overlapped area	1.0	0.93	9.3	0.031
	$A_P = 30.0\ m^2$		$A_v = 5.23$	$P_{K/H} = 0.1743$

Thus vulnerability is enhanced due to secondary kill mode and it further increases due to fire. Results of this section suggest that engine and fuel tank should always be kept apart. This is due to the fact that engine becomes very hot and is always prone to fire if fuel ingestion is there. Now a days aircraft with self sealing materials are available, which can reduce this type of risk.

3.6.2 Redundant Components with no Overlap

So far we have assumed that aircraft has only non-redundant components. It is quite possible that aircraft has redundant components. For illustration purpose, consider an aircraft with two engines. This means if one engine fails, aircraft still can fly with other engine. The kill expression for the aircraft model with redundant components becomes: Aircraft will be killed if pilot is killed or fuel tank is killed or both the engines are killed i.e., kill equation is,

(PILOT) OR (FUEL TANK) OR [ENGINE1) AND (ENGINE 2)]

The equation (3.9) for the probability of aircraft survival given a random hit on it in this case is modified to: Aircraft will be survive if pilot survives and fuel tank survive and one of the engines survives i.e., probability of survival of aircraft is,

(PILOT).AND.(FUELTANK).AND.[(ENGINE1).OR.(ENGINE2)]

That is

$$P_{S/H} = \left(P_{S/H_p} \cdot P_{S/H_f}\ (1 - P_{k/H_{e1}} \cdot P_{k/H_{e2}})\right) \qquad ...(3.13)$$

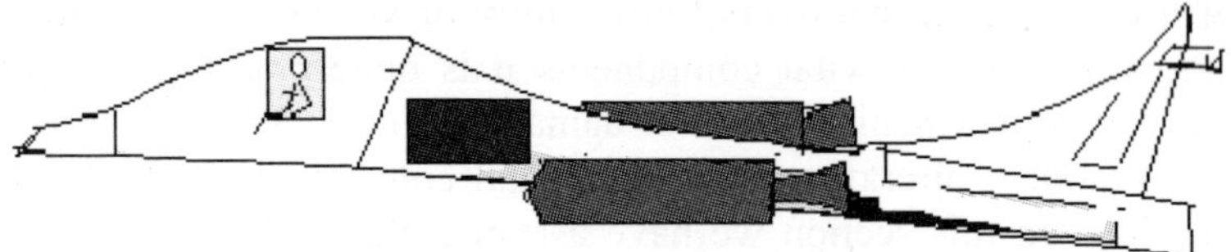

Fig. 3.3: *Redundant components with no overlap.*

If we assume that single hit cannot kill both the engines, then all the components killed are mutually exclusive. Thus from (3.13) one gets:

$$P_{K/H} = P_{k/H_p} + P_{k/H_f} \qquad \text{(From Table 3.1)}$$

$$= 0.0333 + 0.06 = 0.0733$$

It can be seen that by having twin engines, vulnerability of the aircraft has drastically reduced. From 0.1733 it has reduced to 0.0733. In fact for the total kill of aircraft due to engine failure, both the engines have to be killed together.

3.6.3 Redundant Components with Overlap

Now let us consider a case when out of total n non-redundant components c components are overlapping. In this case aircraft will survive only when all the overlapping components survive i.e.,

$$P_{S/h_o} = P_{s1}\, P_{s2} \ldots P_{sc} \qquad \ldots(3.14)$$

Now if out of these c components, two components are redundant (say component number 2 and 3), then aircraft will be killed if both of these are simultaneously killed. If we denote these redundant components as P_{s2} and P_{s3} then equation (3.11)

$$P_{S/H} = (1 - P_{k/H_1})(1 - P_{k/H_2})(1 - P_{k/H_3})$$

is modified as

$$P_{S/h_o} = P_{s1}(1 - P_{s2} \cdot P_{s3}) \ldots P_{s4} \ldots P_{sc} \qquad \ldots(3.14a)$$

As earlier, converting survivability into vulnerability, one can get an equation similar to (3.11).

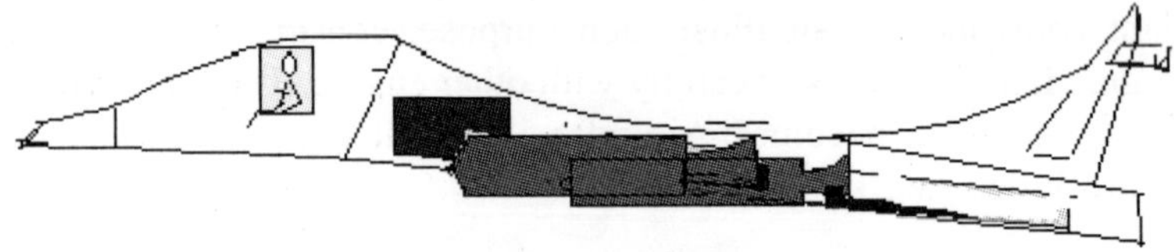

Fig. 3.4: *Redundant component with overlap.*

In this chapter only the basic technique of mathematically evaluating the various cases of aircraft orientation has been discussed. All these cases can occur in a single aircraft when it manoeuvres. We will use these results in sixth chapter and generalise the vulnerability mode.

❑❑❑

DISCRETE SIMULATION

In chapter two, various statistical techniques to be used in mathematical modelling were discussed. These techniques were used in chapter three to model aircraft survivability model. Aim was to give an idea, how one can develop a mathematical model according to a given scenario. There are number of problems which can not be modelled by simple statistical/mathematical techniques or it is very difficult to model by these techniques. Beauty of a model lies, not in its complexity but in its simplicity. For example if one has to make model of a weapon system, techniques of pure mathematics are not so handy and one has to opt for other techniques. In this chapter we will be discussing a different technique which is also quite versatile in solving various problems where events are random. This technique is called Monte Carlo simulation. Computer simulation is one of the most powerful techniques to study various types of problems in system analysis. The conceptual model is the result of the data gathering efforts and is a formulation in one's mind (supplemented by notes and diagrams) of how a particular system operates. Building a simulation model means this conceptual model is converted to a computer model (simulation model). Making this translation requires two important transitions in one's thinking. First, the modeller must be able to think of the system in terms of modelling paradigm supported by the particular modeling software that is being used, second, the different possible ways to model the system should be evaluated to determine the most efficient yet effective way to represent the system. Many problems which can not be solved by mathematical methods can be solved by Monte Carlo simulation.

John von Neumann in the 1940s.

John von Neumann (Neumann János) (December 28, 1903 – February 8, 1957) was a Hungarian mathematician and polymath of Jewish ancestry who made important contributions in quantum physics, functional analysis, set theory, economics, computer science, numerical analysis, hydrodynamics (of explosions), statistics and many other mathematical fields.

Most notably, von Neumann was a pioneer of the modern digital computer and the application of operator theory to quantum mechanics (see Von Neumann algebra), a member of the Manhattan Project Team, and creator of game theory and the concept of cellular automata. Along with Edward Teller and Stamslaw Ulam, von Neumann worked out key steps in the nuclear physics involved in thermonuclear reactions and the hydrogen bomb.

The term "Monte Carlo" was introduced by John von Neumann and Stamslaw Ulam during World War II, as a code word for a secret work at Los Alamos. Monte Carlo is the name of a city in Monaco which is famous for gambling casinos. You know that gambling is basically based on random numbers or throwing of dice, which is also another form of random number generator. Problems dealing with dice were discussed in second chapter. In Monte Carlo technique random numbers are generated and used to simulate random events in the models. Aim of the present chapter will be to discuss various methods of random number generation and apply them to physical problems. Later on, some case studies such as ground target damage, will also be taken up.

Missile has proved its worth in destroying enemy targets with sufficient effectiveness involving lesser cost, as compared to that by aircraft attack. Aircraft, although highly accurate are quite costly and involve enemy attrition rate (i.e., damage of aircraft due to enemy's fire power), whereas missile has a capability of delivering sophisticated heavy warhead loads, with sufficient accuracy, without much attrition rate. Apart from this, risk of human life loss is also involved in the attack by aircraft, which is not present in the case of missile attack (see chapter ten on Cost effectiveness studies). Some of the simple models of damage due to missile attack are discussed in this chapter using Monte Carlo simulation. Aim of the present study is to demonstrate the power of simulation technique and its applications to various practical problems. Case studies like runway and battle field denial by missile warheads will be taken up.

In section 4.1, general introduction and the method of generation of random numbers is discussed alongwith few case studies. Apart from studying the methods of generation of random numbers it is important to study whether these numbers are really random. In section 4.2, methods to test the randomness of numbers are discussed. In section 4.3, an inverse transform method for the generation of random numbers will be studied and in section 4.4, different methods of generating normal random numbers are discussed and few simple case studies will be taken up. In sections 4.5, and 4.6, of this chapter, Monte Carlo random number generation method will be used to simulate the missile attack and to assess the damage incurred to a runway and a battle field.

Generally question is raised, how authentic is the simulation technique? Because of various methods available for the generation of random numbers, one always has a doubt about the authenticity of the method. Each method will have its own errors. Then how to validate the model? Even mathematical model will involve lot of assumptions, which may distract us from ground realities. Another way is to compare the results with available experimental data. All these issues will be studied in this chapter.

4.1 Generation of Uniform Random Numbers

Although the simulation is often viewed as a "method of last resort" to be employed when everything else has failed [53], recent advances in simulation methodologies, availability of software, and technical developments, have made simulation, one of the most widely used and accepted tools in System Analysis and Operations Research. Simulation is also sometimes defined as a technique of performing sampling experiment on the model of the system, using computer.

What are uniform random numbers? To understand this, let us draw a square and divide it into ten equal small squares by drawing horizontal and perpendicular lines. At the centre of each of the small squares, let us put one point. Then we say, these points are uniformly distributed in the bigger square. But these points are not random. By fixing the coordinates of one of the corner of square, we can determine the coordinates of all these points. These points are uniformly distributed but not randomly. Now we generate ten points in the bigger square, by some random number generation technique. If we observe that approximately one point out of these lie in each small square, then in this case we will say points are uniformly distributed and are random, because their location is not fixed and is unknown. When we get into the detailed study of uniform random numbers, we will come to know, it is almost impossible to generate true uniform random points in an area. In the following paragraphs, we will first try to learn, how to generate uniform random numbers between two given numbers. This is one dimensional case and called Monte Carlo method of generating random numbers.

Because the sampling from a particular distribution involves the use of uniform random numbers, stochastic simulation is often called Monte Carlo simulation. Uniform random numbers are the independent random numbers, uniformly distributed over an interval [0,1], and are generally available as a built-in function in most of the computers.

4.1.1 Properties of Random Numbers

A sequence of random numbers has two important statistical properties.

1. Uniformity and,
2. Independence.

Each random number is an independent sample drawn from a continuous uniform distribution between an interval 0 and 1. Probability density function of this distribution is given by,

$$f(x) = \begin{cases} 1, & \text{for } 0 \le x \le 1 \\ 0, & \text{otherwise} \end{cases} \qquad \text{...(4.1)}$$

Reader can see that equation (4.1) is same as equation (2.8) with $a = 0$ and $b = 1$. The expected value of each random number R_i whose distribution is given by (4.1) is given by,

$$E(R) = \int_0^1 x\,dx = \left[\frac{x^2}{2}\right]_0^1 = \frac{1}{2}$$

and variance is given by,

$$\text{Var}(R) = \int_0^1 x^2 dx - [E(R)]^2 = \left[\frac{x^3}{3}\right]_0^1 - \left[\frac{1}{2}\right]^2 = \frac{1}{3} - \frac{1}{4} = \frac{1}{12}$$

Test for the Uniformity of Random Numbers: If the interval between 0 and 1 is divided into n equal intervals and total of m (where $m > n$) random numbers are generated between 0 and 1 then the test for uniformity is that in each of n intervals, approximately (m/n) random numbers will fall.

In this section, we are concerned with the generation of random numbers (uniform) using digital computers. In chapter three, we have given one of the method of generating random numbers, that is dice rolling. With this method we can generate random numbers between two and twelve. Some other conventional methods are coin flipping, card shuffling and roulette wheel. But these are very slow ways of generating random numbers. We can generate thousands of random numbers using computers in no time. Random number hereafter will also be called random variate.

4.1.2 Congruential or Residual Generators

One of the common methods, used for generating the pseudo uniform random numbers is the congruence relationship given by,

$$X_{i+1} \equiv (aX_i + c)(\text{mod } m), \quad i = 1,2,...,n \qquad ...(4.2)$$

where multiplier a, the increment c and modulus m are non-negative integers. Equation (4.2) means, if $(aX_i + c)$ is divided by m, then the remainder is X_{i+1}. In this equation m is a large number such that $m \leq 2^w - 1$, where w is the word length of the computer in use for generating the $(m- 1)$ numbers and $(i = 0)$ is seed value. By seed value, we mean any initial value used for generating a set of random numbers. Seed value should be different for different set of random numbers. In order, the numbers falling between 0 and 1, we must divide all X_is' by $(m- 1)$. To illustrate equation (4.2), let us take, $a = 3$, $X_0 = 5$, $c = 3$ and $m = 7$. Then

(*i*) $X_1 \equiv (3 \times 5 + 3) \text{ (mod 7)} = 4$
(*ii*) $X_2 \equiv (3 \times 4 + 3) \text{ (mod 7)} = 1$
(*iii*) $X_3 \equiv (3 \times 1 + 3) \text{ (mod 7)} = 6$
(*iv*) $X_4 \equiv (3 \times 6 + 3) \text{ (mod 7)} = 0$
(*v*) $X_5 \equiv (3 \times 0 + 3) \text{ (mod 7)} = 3$
(*vi*) $X_6 \equiv (3 \times 3 + 3) \text{ (mod 7)} = 5$
(*vii*) $X_7 \equiv (3 \times 5 + 3) \text{ (mod 7)} = 4$

Thus we see that numbers generated are, 4, 1, 6, 0, 3, 5, 4. Thus there are only six non-repeating numbers for $m = 7$. Larger is m, more are the non-repeated numbers. Thus period of these set of numbers is m. There is a possibility that these numbers may repeat before the period m is achieved. Let in the above example $m = 9$, then we see that number generated are 0, 3, 3, 3, 3,... This means after second number it starts repeating. It has been shown [53] that in order to have non-repeated period m, following conditions are to be satisfied,

(*i*) c is relatively prime to m, i.e., c and m have no common divisor.
(*ii*) $a \equiv 1 \text{ (mod } g)$ for every prime factor g of m.
(*iii*) $a \equiv 1 \text{ (mod 4)}$ if m is a multiple of 4.

Condition (*i*) is obvious whereas condition (*ii*) means $a = g\{a/g\} + 1$, where number inside the bracket { } is integer value of a/g. Let g be the prime factor of m; then if $\{a/g\} = k$, then we can write

$$a = 1 + gk$$

Condition (*iii*) means that

$$a = 1 + 4\{a/4\}$$

if $m/4$ is an integer.

Based on these conditions we observe that in the above example $m = 9$ had a common factor with a, thus it did not give full period of numbers. Name pseudo random numbers, is given to these random numbers. Literal meaning of pseudo is false. They are called pseudo because to generate them, some known arithmetic operation is used, which can generate non-recurring numbers but they may not be truly uniformly random.

Thus before employing any random number generator, it should be properly validated by testing the random numbers for their randomness. For testing the randomness of random numbers, some tests have been given in the coming sections.

4.1.3 Computation of Irregular Area using Monte Carlo Simulation

To further understand Monte Carlo simulation, let us examine a simple problem. Below is a rectangle for which we know the length [10 units] and height [4 units]. It is split into two sections which are identified using different colours. What is the area covered by the black colour?

Due to the irregular way in which the rectangle is split, this problem cannot be easily solved using analytical methods. However, we can use Monte Carlo simulation to easily find an approximate answer. The procedure is as follows:

1. Randomly select a location (point) within the rectangle.
2. If it is within the black area, record this instance a hit.
3. Generate a new location and follow 2.
4. Repeat this process 10,000 times.

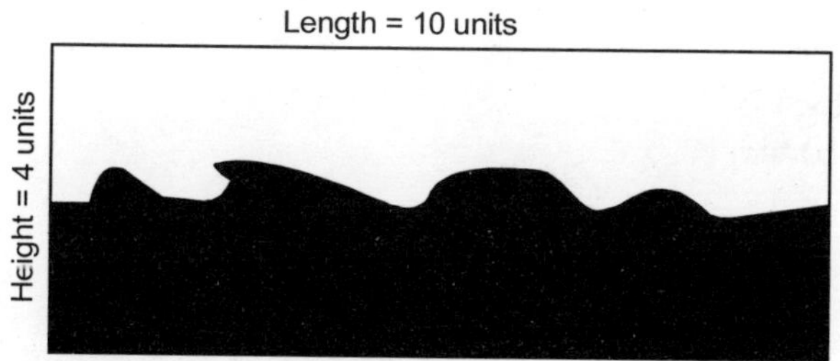

What is the area covered by Black?

After using Monte Carlo simulation to test 10,000 random points, we will have a pretty good average of how often the randomly selected location falls within the black area. We also know from basic mathematics that the area of the rectangle is 40 square units [length × height]. Thus, the black area can now be calculated by:

$$\text{Black area} = \frac{\text{number of black hits}}{10{,}000 \text{ hits}} \times 40 \text{ square units}$$

Naturally, the number of random hits used by the Monte Carlo simulation doesn't have to be 10,000. If more points are used, an even more accurate approximation for the black area can be obtained.

4.1.4 Multiplicative Generator Method

Another widely used method is multiplicative generator method and is given as,

$$X_{i+1} \equiv (aX_i)(\text{mod } m), \quad i = 1, 2, ..., n \qquad ...(4.3)$$

This equation is obtained from (4.2) by putting $c = 0$. One important condition in this is that X_0 is prime to m and a satisfies certain congruence conditions. Various tests for checking independence and uniformity of these pseudo random numbers are given in [53]. History of random number generators is given in [86]. In this case too, in order to generate random numbers between 0 and 1, we divide X_{i+1} by m.

Below, we are given a *C++* program for generating the random numbers by Congruential method.

Program 4.1: Congruential Method

```
/*program for generating uniform random numbers by Congruential method*/
#include <iostream.h>
```

```
#include<stdlib.h>
#include <math.h>
main( )
{
 int a,b,k,j,i,m,nn,seed,r[50];
cout<<"\nEnter the integer value of a,b,m, leave space between a,b,m";
cin>>a>>b>>m;
cout<<"\nEnter the integer value of seed";
cin>>seed;
cout<<"\nEnter the number of Random numbers to be generated";
cin>>nn;
r[0]=seed;
for(i=0; i<=nn; ++i)
{
r[i]=(a*r[i-1]+b)%m;
cout<<r[i];
}
/* Program ends*/
}
```

4.1.5 Mid Square Random Number Generator

This is one of the earliest method for generating the random numbers. This was used in 1950s, when the principle use of simulation was in designing thermonuclear weapons. Method is as follows:

1. Take some *n* digit number.
2. Square the number and select *n* digit number from the middle of the square number.
3. Square again this number and repeat the process.

Example 4.1: Generate random numbers using Mid Square Random Number Generator.

Solution: Let us assume a three digit seed value as 123.

Step 1: Square of 123 is 15129. We select mid three numbers which is 512.

Step 2: Square of 512 is 262144. We select mid two numbers which is 21.

Repeat the process. Thus random numbers are 512, 21, …

Number so-obtained is a desired random number. A computer program in C++ for generating random number by mid-square method is given below.

Program 4.2: Mid Square Random Number Generator

```
/* Generation of Random numbers by midsquare method*/
#include <iostream.h>
#include<stdlib.h>
#include <math.h>
main( )
```

```
{
 long int i,s,z1,x,nd,seed;
int n;
float z,y;
cout<<"\n Give the seed number of four digits";
cin>>seed;
cout<<"\n Give the number of random numbers to be generated";
cin>>n;
for(i=0; i<=n; ++i)
{
y=int(seed*seed/100.0);
//cout<<y<<endl;
z=(y/10000.0);
z1=int(z);
x=int((z-z1)*10000);
seed=x;
cout<<x<<"\n";
}
}
```

OUTPUT OF PROGRAM

A sequence of 25 random numbers with seed value 3459 is given below.

9645	0260	0675	4555	7479	9353	4785	8962	3173
678	4596	1231	5153	5534	0250	0625	3905	2489
1950	8025	4005	0400	1599	5567	9913	2675	

4.1.6 Random Walk Problem

One of the important applications of random numbers is a drunkard walk. A drunkard is trying to go in a direction (say y-axis in xy-plane). But sometimes he moves in forward direction and some times left, right or backward direction. Random walk has many applications in the field of Physics. Brownian motion of molecules is like random walk. Probabilities of drunkard's steps are given as follows.

Probability of moving forward = 0.5

Probability of moving backward = 0.1

Probability of moving right = 0.2

Probability of moving left = 0.2

Program 4.3: Random Walk Program

```
//C++ program for generating Random Walk of a drunkard.
#include <iostream.h>
#include <stdlib.h>
```

```
#include <time.h>
main( )
{
 int x=0, y=0, t=0;
//Parameters have been initialised at definition level.
while(y<10)
{rand( );
if(rand( )>=0 ||rand( )<=4)
 y++;
else if(rand( )=5 )
y--;
else if(rand( )>=6 ||rand( )<=7)
x++;
else
x--;
}t++;
cout<<"t="<<t<<"x="<<x<<"y="<<y;
}}
```

Table 4.1: Random walk

Step no.	Random number	Movement	x-coordinate	y-coordinate
1	6	R	1	0
2	1	F	1	1
3	0	F	1	2
4	1	F	1	3
5	7	R	2	3
6	0	F	2	4
7	6	R	3	4
8	6	R	4	4
9	7	R	5	4
10	4	F	5	3
11	7	R	6	3
12	2	F	6	4
13	1	F	6	5
14	5	B	5	5
15	3	F	5	6

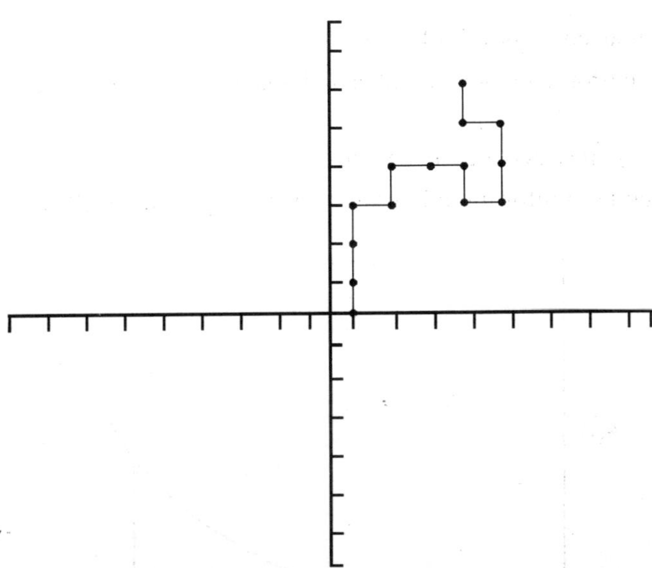

Fig. 4.1: *Drunkard's random walk.*

It is assumed that he takes one step per minute and his destination is ten steps in the direction of y-axis. Find out the time taken by the drunkard to reach the destination. In order to solve this problem, we generate ten uniform random numbers, between 0 and 9 using Congruential method. If number lies between 0 and 4, y-coordinate is incremented by one step. If number is 5, y-coordinate is decremented by one step. If number is 6 or 7, x-coordinate is incremented by one step and if number is 8 or 9, x-coordinate is decremented by one step. Algorithm of this logic is given in C++ program 4.3 and results are given in Fig. 4.1.

4.1.7 Acceptance Rejection Method of Random Number Generation

This method is sometimes called, *rejection method.* This method is used for generating random numbers from a given non-uniform distribution. Basically this method works by generating uniform random numbers repeatedly, and accepting only those that meet certain conditions. These conditions for accepting the uniform random numbers are so designed that the accepted random numbers follow the given distribution. For the rejection method to be applicable, the probability density function $f(x)$ of the distribution must be non zero over an interval, say (a, b). Let function $f(x)$ is bounded by the upper limit f_{max} which is the maxima of $f(x)$.

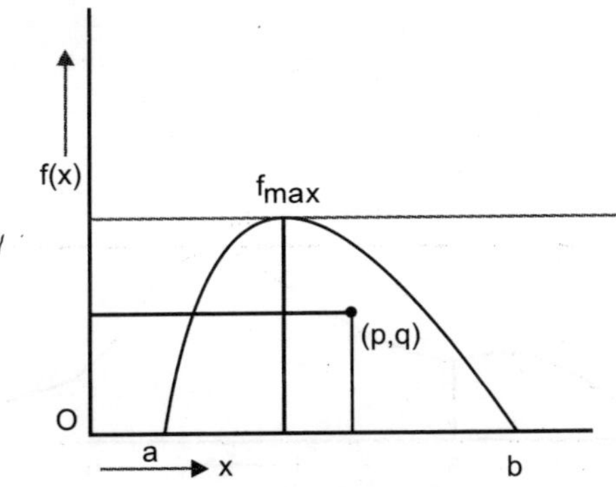

Fig. 4.2: *Rejection method.*

The rejection method consists of following steps:

1. Generate a uniform random number u lying between (0, 1). Let us define p as

   $$p = a + (b - a)u$$

   This means p lies between a and b.

2. Generate another uniform random number v lying between (0, 1). Let us define $q = f_{max} \cdot v$.

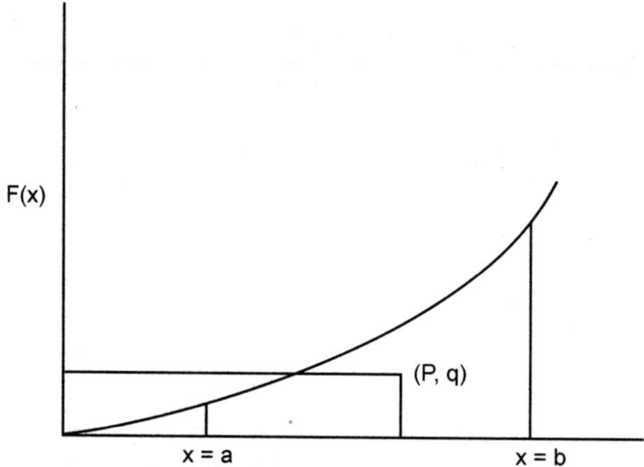

Fig. 4.3: *Rejection method for curve having no maxima.*

3. If $q > f(y)$, then reject the pair (u, v), otherwise accept u as the random number following the distribution $f(x)$.

This method works as all the random numbers accepted lie below the curve $y = f(x)$. Only restriction is that this method works for random number in a limited area, which is bounded by lower and upper limits. In this method one has to generate large number of uniform random numbers which may take more time in generating the desired random numbers.

In case curve $f(x)$ does not have a maxima *viz.*, curve is concave upward, then above step no.2 becomes $q = f(b) \cdot v$, rest of the procedure is same (Fig. 4.3).

In case there are more than one cusp in the probability distribution function, highest of maximum values can be taken as f_{max} (see Fig. 4.4).

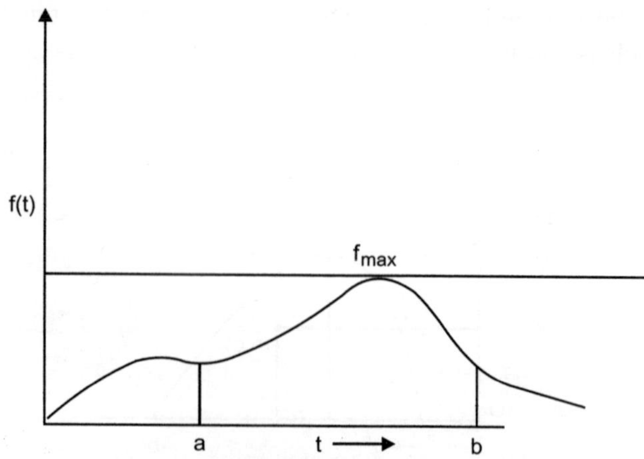

Fig. 4.4: *Probability density function with two maxima.*

4.1.8 Which are the Good Random Numbers?

- It should have a sufficiently long cycle, without repetition.
- A set of random numbers should be able to repeat. This means, given the same parameters and seed value, set of random numbers should be same.
- Generated random numbers should be independent and uniform.
- Random number generator should be fast and cost effective.
- It should be platform independent.

4.2 Testing of Random Numbers

To find out whether a given series of random numbers are truly random, there are several tests available. Random numbers are considered random if

- The numbers are uniformly distributed i.e., every number has equal chance to occur.
- The numbers are not serially autocorrelated.

Meaning of second point is that once a random number is generated, next can not be generated by some correlation with one. Based on the above specifications, to test whether given random numbers are random or not, below we are give some tests.

4.2.1 The Kolmogrov-Smirnov Test

The test compares the continuous CDF (Cumulative Distribution Function has been discussed in section 2.5.1), $F(x)$ of the uniform distribution with the empirical CDF(Appendix 4.5), $S_N(x)$, of the sample of N random numbers. The largest absolute deviation between $F(x)$ and $S_N(x)$ is determined and is compared with the critical value, which is available as function of N in Kolmogrov-Smirnov tables (given at the end of this chapter as Appendix-4.4), for various levels of significance. Below we give an example, to explain Kolmogrov-Smirnov Test.

Example 4.2: Ten random numbers are given as follows:

0.26, 0.88, 0.12, 0.52, 0.23, 0.43, 0.51, 0.66, 0.79, 0.65

We have to test the uniformity of these numbers with a level of significance of $\alpha = 0.5$. In Table 4.2, first row shows the random numbers, second shows empirical distribution i.e., i/N, third gives their difference (the maximum of which is D^+ say) and last row gives deviation $R_i - (i - 1)/N$ (the maximum of which is D^-).

Table 4.2: Kolmogrov-Smirnov test of uniform random numbers

R_i	0.12	0.23	0.26	0.43	0.51	0.52	0.65	0.66	0.79	0.88
i/N	0.10	0.20	0.30	0.40	0.50	0.60	0.70	0.80	0.90	1.0
$i/N-R_i$	0.02	0.03	0.04	0.03	0.01	0.08	0.05	0.14	0.11	0.12
$R_i-(i-1)/N$	0.26	0.13	0.06	0.13	0.11	0.05	0.05	0.04	0.01	0.02

Let us denote by D, the maximum of D^+ and D^-. The critical value of D from Kolmogrov-Smirnov tables for $\alpha = 0.5$, and $N = 10$ is 0.410. Value of D from the Table 4.2 is max(0.14, 0.26) i.e., 0.26 which is less than 0.41. Hence these random numbers are uniform with level of significance $\alpha = 0.5$.

4.2.2 chi-square (χ^2) Test

Another popular test for testing the uniformity level of random numbers is known as chi-squarc (χ^2) test. The chi-square distribution is a special form of Gamma distribution. The chi-square test provides a useful test of goodness of fit, that is, how well data from an empirical distribution of n observations conform to the model of random sampling from a particular theoretical distribution. If there were only two categories say, success and failure, the model of independent trials with probability p of success is tested using the normal approximation to the binomial distribution. But for data in several categories, the problem is how to combine the test for different categories in a reasonable way. This problem was solved as follows, by the statistician Karl Pearson (1857–1936). For a finite number of categories m, let N_i denotes the number of results in category i. Under the hypothesis that the N_i are counting results of independent trials with probability p_i, for large enough n the so-called chi-square statistics

$$\sum_{i=1}^{m} \frac{(N_i - np_i)^2}{np_i}$$

that is the sum over categories of (observed – expected)^2/expected, has distribution that is approximately chi-square with $m-1$ degree of freedom. In statistical jargon, a value of the statistic higher than the 95th percentile point on the chi-square distribution with $m-1$ degree of freedom would "reject the hypothesis at the 5% level".

Thus the chi-square test uses the sample statistic

$$\chi^2 = \sum_{i=1}^{n} \frac{(O_i - E_i)^2}{E_i} \qquad ...(4.4)$$

where $O_i = N_i$ is the observed random numbers in i-th class, E_i is the expected number in the i-th class and n is the number of classes. For the uniform distribution E_i, the expected number in each class is given by

$$E_i = \frac{N}{n}$$

for equally spaced n classes, where N is the total number of observations.

Table 4.3: chi-square test

Class	Random numbers falling i-th class	$O_i - E_i$	$(O_i - E_i)^2$
0.1<rn≤.2	11	1	1
0.2<rn≤.3	8	2	4
0.3<rn≤.4	9	1	1
0.4<rn≤.5	12	2	4
0.5<rn≤.6	13	3	9
0.6<rn≤.7	12	2	4
0.7<rn≤.8	12	2	4
0.8<rn≤.9	8	2	4
0.9<rn≤1.0	9	1	1
			32

It can be shown that the sampling distribution of χ^2 is approximately the chi-square distribution with (n – 1) degree of freedom. To make the process clearer, let us generate hundred rn numbers of uniform random numbers lying between 0 and 1. Divide these numbers in ten classes (n) of equal interval so that random numbers less than or equal to 0.1 fall in the 1st class, those of $0.1 < rn \leq .2$ fall in 2nd class, $0.2 < rn \leq .3$ fall in 3rd class and so on. In this way, let us assume O_i number of random numbers fall in the i-th class. E_i here is 100/10 = 10. Table 4.2 gives O_i in different classes in second column. Third column gives the difference $O_i - E_i$ and fourth column square of $O_i - E_i$. Then using equation (4.4) we get,

$$\chi^2 = 32/10 = 3.2$$

From the χ^2 tables in the Appendix 4.3, we find that for degree of freedom 9, value of χ^2 for 95% level of confidence is 16.919, which is more than our value 3.2. Thus random numbers of Table 4.2 are uniform with 95% level of confidence.

4.2.3 Poker's Method

This test is named after a game of cards called poker. In this game five cards are distributed to each player out of pack of fifty two cards. The cards are ranked from high to low in the following order: Ace, King, Queen, Jack, 10, 9, 8, 7, 6, 5, 4, 3, 2. Aces are always high. Aces are worth more than Kings which are worth more than Queens which are worth more than Jack, and so on. Each player is dealt five cards. The object of the game is to end up with the highest-valued hand. From best to worst, hands are ranked in the following order:

1. **Royal Flush:** A Royal Flush is composed of 10, Jack, Queen, King and Ace, all of the same suit.
2. **Straight Flush:** Comprised of five cards in numerical order, all of the same suit.
3. **Four of a Kind:** Four cards of the same numerical rank and another random card.
4. **Full House:** Of the five cards in one's hand, three have the same numerical rank, and the two remaining card also have the same numerical rank.
5. **Flush:** A Flush is comprised of five cards of the same suit, regardless of their numerical rank. In a tie, whoever has the highest ranking card wins.
6. **Straight:** Five cards in numerical order, regardless of their suits.
7. **Three of a Kind:** Three cards of the same numerical rank, and two random cards that are not a pair.
8. **Two Pair:** Two sets of pairs, and another random card.
9. **One Pair:** One pair and three random cards.

High Card

Poker test not only tests the randomness of the sequence of numbers, but also the digits comprising of each number. Every random number of five digits or every sequence of five digits is treated as a poker hand. For example

13586 are five different digits (Flush)

44138 would be a pair

22779 would be two pairs

33328 would be three of kind

55588 would be a full house

55559 would be four of kind

77777 would be five of kind

The occurrence of five of kind is rare. The order of the 'cards' within a 'hand' is unimportant in the test. The straight, flushes, and royals of the Poker are disregarded in the poker test.

In 10,000 random and independent numbers of five digit each, one may expect the following distribution of various combinations.

Table 4.4: Actual outcome of Poker's combinations

Five different digits	3024	30.24%
Pairs	5040	50.40%
Two-pairs	1080	10.80%
Three of a kind	720	7.20%
Full houses	90	0.90%
Four of a kind	45	0.45%
Five of a kind	1	0.01%

Number of outcomes in column two of the above table are taken as expected values E_i as in chi-square test. Exact outcome is taken as O_i. Like chi-square test standard deviation is computed.

Example 4.3: Poker Test

A sequence of 10000 five digits numbers are generated. Outcome of different categories of combination is as given in Table 4.5.

Table 4.5: Poker's test

Combination distribution	*Observed distribution*	*Expected values $(O_i - E_i)$*	$\frac{(O_i - E_i)^2}{E_i}$
Five different digits	3033	3024	0.0268
Pairs	4932	5040	2.3143
Two-pairs	1096	1080	0.2370
Three of a kind	780	720	5.0
Full houses	99	90	0.9
Four of a kind	59	45	4.355
Five of a kind	01	01	0.0
	10,000	10,000	12.8336

The degrees of freedom in this case are six, one less than the number of combinations. The critical value of χ^2 for six degree of freedom at a = 0.01 is 16.8. The value obtained in the last column is 12.8336, which is less than the critical value. Hence, the number generated are independent.

4.2.4 Testing for Auto Correlation

In section 4.2.2, we tested the uniformity of random numbers using chi-square test. But uniformity is essential condition of random numbers, not the sufficient condition. For example if numbers .1, .2, .3, .4, ... are tested by chi-square test, they will be hundred percent uniform with chi-square value equal to zero, but numbers have correlation. This means they are not random. This means we need another test for testing the correlation of the random numbers. We device another test so that correlation between two adjoining random numbers is tested i.e., pair of random numbers are to be taken. We check the random numbers in two dimensions. We construct a square matrix say of $n \times n$ cells. For simplicity we take $n = 3$. In this way rows and columns of matrix have size 0.33. Cell(i,j) of the matrix are shown in Fig. 4.5. We chose first two numbers, say R_1 and R_2. If $R_1 \leq 0.33$ and $R_2 \leq 0.33$, then both numbers lie in cell (1,1) of the matrix. Next we chose second and third number. Same way we test that in which cell both numbers fall. This process is continued till all the succeeding pairs are over. Then we count random pairs in each cell. Expected numbers of pairs in each cell are:

$$E_i = (\text{total number of random numbers} - 1)/\text{number of cells}$$

This way frequency O_i of pair in each cell is calculated. Thus chi-square is computed as in section 4.2.2. This concept is further explained in the following example.

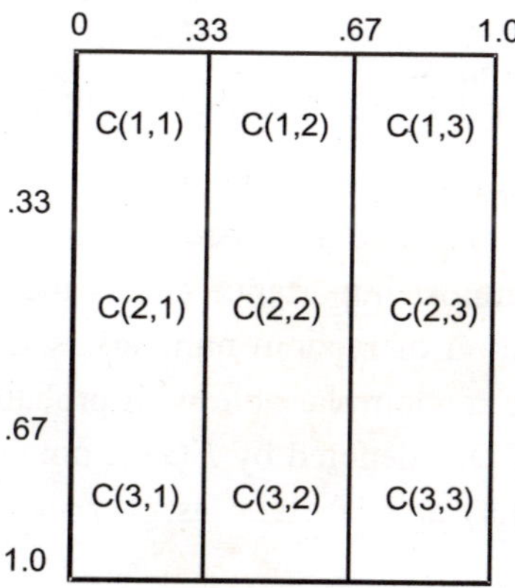

Fig. 4.5: *Distribution of random numbers in cells.*

Example 4.4: Following random numbers have been generated by congruential method. Test their randomness.

49 95 82 19 41 31 12 53 62 40 87 83 26 01 91 55 38 75 90 35 71 57 27 85
52 08 35 57 88 38 77 86 29 18 09 96 58 22 08 93 85 45 79 68 20 11 78 93
21 13 06 32 63 79 54 67 35 18 81 40 62 13 76 74 76 45 29 36 80 78 95 25 52.

Solution: These are 73 numbers and we divide them into 72 pair such that 49, 95, is first pair, 95, 82 is second pair, 82, 19 is third pair and so on. If we call each pair as R_1 and R_2 then condition is $R_1 \leq .67$ & $R_2 \leq 1.0$ thus it goes to cell $C(2, 3)$. Similarly we place other pairs in their respective cells. Following table gives the frequency of pairs in each cell.

Cell	Frequency (O_i)	($O_i - E_i$)	$(O_i - E_i)^2$
C(1,1)	9	1	1
C(1, 2)	7	1	1
C(1, 3)	6	2	4
C(2, 1)	6	2	4
C(2, 2)	8	0	0
C(2, 3)	9	1	1
C(3, 1)	7	1	1
C(3, 2)	9	1	1
C(3, 3)	11	3	9
Sum	72		24

Therefore

$$\chi^2 = 24/8 = 3.0$$

In this case since there are two variable R_1 and R_2 and hence the degree of freedom is nine minus two i.e., seven. The criterion value of χ^2 for seven degree of freedom at 95% level of confidence is 14.067. This value is much higher than the values obtained by present test and hence given random numbers are not auto correlated.

4.3 Random Variate For Non-uniform Distribution

In above section some basic techniques of generating uniform random numbers were briefly explained. Random number generation in itself is a field and needs a full book. Generation of random numbers as per the given distribution is of utmost importance and is the subject of this section. One of the important techniques for the generation of random numbers is **Inverse Transform Method**, which will be discussed here. Let X be the random variable with probability distribution function $f(x)$ and Cumulative Distribution Function (CDF) denoted by $F(x)$. Cumulative distribution function has been studied in chapter two (section 2.4.4) and is given as,

$$F(x) = \int_{-\infty}^{x} f(x)dx \qquad ...(4.5)$$

where $0 \leq F(x) \leq 1$, since the integral of probability function $f(x)$ over values of x varying from $-\infty \leq f(x) \leq +\infty$ is 1. You can observe some similarity between random numbers generated in previous section and CDF. Can you guess what? Both vary between 0 and 1. Figure 4.5 shows a typical cumulative distribution function.

The cumulative function can be solved for x; i.e., if $y = F(x)$, then

$$x = F^{-1}(y) \qquad ...(4.6)$$

Now we will prove that if y is uniformly distributed over a region $0 \leq y \leq 1$ then variate X, has a distribution whose values x, are given by equation (4.5).

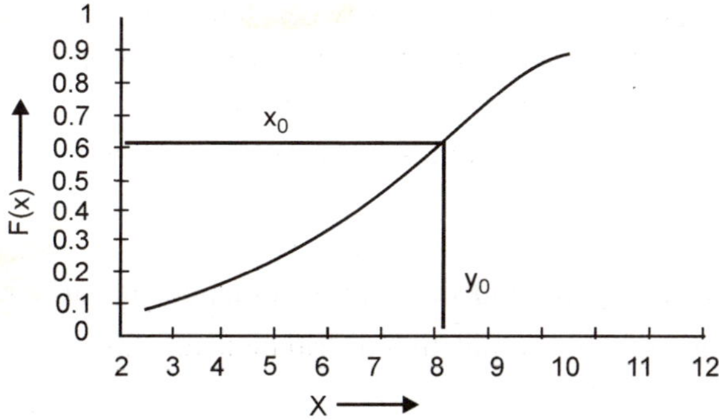

Fig. 4.6: *Variation CDF versus X.*

If X is a random variate governed by the given probability density function $f(x)$, and that $Y = F(x)$ is a corresponding value of the cumulative distribution, then

$$P\{Y \le y_0\} = P\{X \le x_0\}$$

But by the definition of cumulative density function, $P\{X \le x_0\} = F\{x_0\} = y_0$, hence

$$P\{Y \le y_0\} = y_0 \qquad \text{...(4.7)}$$

which is the expression for the cumulative uniform distribution within the interval (0,1). Thus we see that Y is uniformly distributed in the interval (0,1), irrespective of the distribution of X. This fact is also clearly shown in Fig. 4.6.

Therefore in order to generate a random variate X with distribution $f(x)$, we first generate a uniform random variate U lying between (0,1). Then we obtain X as,

$$X = F^{-1}(U)$$

It is often observed that it is not possible to integrate the function $f(x)$ to get the CDF (normal distribution function is one of the example). Some time it is possible to get CDF of $f(x)$ but is not possible to invert it. Thus inverse transform method fails. In such cases we have to adopt alternative methods. There are other methods for the generation of random numbers, for example, composition method and acceptance-rejection method. Since study of basic theory of random number generation is not the aim of the book, readers may see reference [53, 77, 86] for detailed study. A typical example of determination of required random number is given below.

Example 4.5: Generate a random variable with uniform distribution $f(x)$ given by,

$$F(x) = \begin{cases} \dfrac{1}{b-a}, & a \le x \le b \\ 0, & \text{otherwise} \end{cases}$$

Solution: Cumulative distribution function is given by,

$$F(x) = \begin{cases} 0, & x < a, \\ \dfrac{x-a}{b-a}, & a \le x \le b \\ 1, & x > b \end{cases}$$

and thus solving for $x = X$, one gets,

$$X = F^{-1}(U) = a + (b - a)U \qquad \text{...(4.8)}$$

Thus X is a uniform random variable lying between limits a and b. In the following example, we illustrate a problem which is reverse of the problem given in example 4.5 i.e., when distribution of random variate is given, determine the distribution function.

Example 4.6: Generate a distribution function $f(x)$ when a random variable X as a function of uniform distribution is given.

Solution : Let us consider a random variable given by $X = -2\ln(U)$, where U is a uniform random variable between [0,1]. Our aim is to find distribution function of variable X.

Now by definition (equation 4.3), cumulative distribution function of random variable X is given by

$$F_X(x) = P(X \le x)$$

or

$$P(X \le x) = P(X \le -2\ln(U) \le x) = P(2\ln(U) \ge -x) = P(\ln(U) \ge -x/2)$$

$$= P(e^{\ln(U)} \ge e^{-x/2})$$

$$= P(U \le e^{-x/2}) = \int_{e^{-x/2}}^{\infty} f_U(u) \cdot du$$

where $f_U(u)$ is the probability density function of a random variable that is uniform on [0,1]

If we assume that $f_U(u) = 1$ on [0,1] and 0 elsewhere, we find that

$$F_X(x) = \begin{cases} \int_{e^{-x/2}}^{1} 1\,du + \int_{1}^{\infty} 0\,du = 1 - e^{-x/2}, & \text{if } x \ge 0 \\ \int_{e^{-x/2}}^{\infty} 0\,du = 0, & \text{if } x < 0 \end{cases}$$

Therefore the probability density function of X is

$$F_X(x) = \frac{d}{dx} F_X(x) = \frac{1}{2} e^{-x/2}$$

if $x \ge 0$ and 0 otherwise. This is the well-known exponential distribution with mean $E(X) = 2$.

4.4 Normal Random Number Generator

In the previous section we have given one of the method i.e., inverse transform method, for the generation of non-uniform random numbers. Normal random numbers, which follow normal distribution, have application in various problems. Distribution of marks obtained by a class of students, follows normal distribution. Similarly hit points, fired by a gun follow normal distribution. Thus for target damage problems, normal random numbers are often needed. In this section we will discuss few techniques of normal random number generation.

Suppose we wish to generate random numbers X_1, X_2, ..., which are independent and have the normal distribution with mean M and variance σ^2. Such normal numbers are often denoted by $N(\mu, \sigma^2)$. The cumulative distribution function of normal distribution $N(\mu, \sigma^2)$ is given by

$$F_X(x) = \frac{1}{\sqrt{2\pi\sigma}} \int_{-\infty}^{x} e^{-(z-\mu)^2/2\sigma^2} dz$$

We know that there is no closed-form expression for $F_X(x)$, and also it is not possible to solve this expression for random variable X.

One of the approximation for $F_X(x)$ is given by Shah (see reference [18]), which states that, for $\mu = 0$ and $\sigma = 1$,

$$F_X(x) = x(4.4 - x)/10 + 0.5 \qquad ...(4.9)$$

with an error of the order 0.005 for x lying between 0 and 2.2. This method is however not recommended for the generation of normal random numbers.

Algorithm that generates $N(0,1)$ random variables can always be modified to become general $N(\mu, \sigma^2)$ generators, as follows:

$$N(\mu,\sigma^2) = \mu + \sigma N(0,1) \qquad ...(4.10)$$

In the following sections few more accurate methods of normal random number generations will be discussed.

4.4.1 Central Limit Theorem Approach

One of the important but not very accurate methods of generating normal random numbers is given by Central Limit Theorem. This theorem can be stated as follows;

If $X_1, X_2, ..., X_n$ are independently distributed uniform random variables according to some common distribution function with mean μ and finite variance σ^2, then as the number of random variables increase indefinitely, the random variable

$$Y = \frac{\sum_{i=1}^{n} X_i - n\mu}{\sqrt{n\sigma^2}} \qquad ...(4.11)$$

converges to a normal random variable with mean 0 and variance 1.

Now expected value and variance of a uniform random number U on $[a\ b]$ with $a < b$ is,

$$E(U) = (a + b)/2 \text{ and } \mathrm{Var}(U) = (b - a)^2/12$$

Thus if $a = 0$ and $b = 1$, we will have

$$E(U) = \frac{1}{2} \text{ and Var} = 1/12$$

Thus in equation (4.11) if X_i is a uniform random variable on [0, 1] then

$$Y = \frac{\sum_{i=1}^{k} U_i - k/2}{\sqrt{k/12}} \qquad ...(4.12)$$

has mean 0 and variance 1. The distribution function of Y is close to that of the standard normal distribution for large k. In fact in equation (4.11), k need not be very large. According to Stirling's approximation $k = 12$ is sufficient to generate normal number accurately.

4.4.2 Box-Muller Transformation

Another method of generating normal random variables is due to Box, CEP and Muller, ME (1958)[9], which states that if U_1 and U_2 are independent uniform random variables on the interval [0,1], then (see appendix 4.1)

$$X_1 = \sqrt{-2\ln(U_1)}\ \sin(2\pi U_2),$$

is exactly $N(0,1)$.

Using this method, one can generate a sequence of independent $N(\mu, \sigma^2)$ random variables to be

$$\left.\begin{array}{ll} \mu+\sigma X_1, & X_1 = \sqrt{-2\ln(U_1)}\sin(2\pi U_2), \\ \mu+\sigma X_2, & X_2 = \sqrt{-2\ln(U_3)}\sin(2\pi U_4), \\ \mu+\sigma X_3, & X_3 = \sqrt{-2\ln(U_5)}\sin(2\pi U_6), \end{array}\right\} \quad \text{...(4.13)}$$

This method is relatively fast, and provides exact normal random variables. Sometimes cos (.) in place of sin (.) is also used. This also gives normal random numbers different from those given by sin (.) and is independent of each other. Care must be taken that all *U*s' must be independently chosen from different streams of independent numbers.

4.4.3 Marsaglia and Bray Method

Marsaglia and Bray [37,59] modified the Box-Muller method to avoid the use of trigonometric functions since the computation of these functions is likely to be relatively slow. Their method is:

(*i*) Generate two $U(0,1)$ random variables u_1, u_2.

(*ii*) Calculate $w_1 = 2u_1 - 1$, $w_2 = 2u_2 - 1$

(*iii*) Calculate $w = w_1^2 + w_2^2$; if $w \geq 1$, return to (1)

(*iv*) Calculate $c = (-2\log w_1 / w_2)^{1/2}$, $z_1 = c * w_1$, $z_2 = c * w_2$

This method may be a bit faster but gives results similar to Box-Muller method.

4.5 Applications of Random Numbers

Uniform random number generation can be used to solve various types of problems in defence and industries. An interesting application of uniform random numbers have already given in section 4.1.2, where movement of a drunkard was recorded. Just to illustrate the power of simulation, we will demonstrate one interesting application, for which although simulation technique is never used yet it has an academic interest. This application is the evaluation of an integral, which otherwise can not be integrated by analytical methods. Let such integral be,

$$I = \int_a^b f(x)dx \quad \text{...(4.14)}$$

subject to the condition $0 \leq f(x) \leq f_{max}$, where f_{max} = maximum value of $f(x)$ in the range $a \leq x \leq b$.

To illustrate this method, let us assume, $f(x) = \frac{1}{\sqrt{2\pi\sigma^2}} e^{-x^2/2\sigma^2}$, $a = -5$ and $b = 1$. It can be seen that this is a normal distribution function with

$$f_{max} = \frac{1}{\sqrt{2\pi\sigma^2}}$$

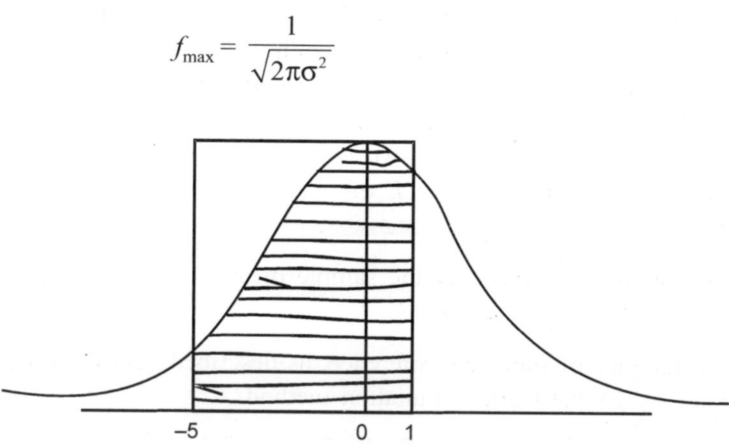

Fig. 4.7: *Shaded portion is the value of integral.*

For the case of simplicity let us assume that $\sigma = 1$ in this integral.

Now generate a uniform random number X having value x, such that $-5 \le x \le 1$(see example 4.1 for the generation of uniform random number between the limits a and b). Then determine y from the relation $y = f(x)$ for this value of X and call it y_{max}. Generate another uniform random number Y having value y, so that $0 \le y \le f_{max}$. If point (x,y) falls below the curve $y = f(x)$, i.e., $y \le y_{max}$, count this point otherwise reject. Repeat this process N number of times, and generate N points where N is a large number. Out of these N trials if n number of times, condition $y = f(x)$ ($y \le y_{max}$) is satisfied then value of integral by Mean Value Theorem is,

$$I = \frac{n}{N}(b-a)\cdot f_{max} \qquad \text{...(4.15)}$$

It is seen that when $N = 8693$, value of integral obtained from (4.15) is 0.8635 whereas exact value of the integral by numerical methods is 0.8414, if $N = 84352$, value of integral becomes 0.8435, which is quite close to numerical value. The error between the two values further will decrease if N is increased. Simulation method is never used for integration purpose, as accurate numerical techniques are available and it takes quite less time. But simulation technique is quite versatile for various other physical situations, which cannot be modelled otherwise, or are quite difficult to model mathematically. Some of these situations will be discussed in the following sections.

4.5.1 Damage Assessment by Monte Carlo Method

Target damage is one of the important problems in warfare. There are cases in target damage studies, where simulation by Monte Carlo method of random number generation is the only techniques by which it can be handled. To understand this, first we will solve a simple problem of coverage of target, by simulation.

Consider a square ground target say, T. Let E represents the total coverage of T due to various warheads including overlapping, if any; and D the overlapping coverage. We assume that center of the target is the aim point and three bombs are dropped at its center (Fig. 4.8). To estimate the expected value of E and D, one generates a stream of random numbers from appropriate normal distributions to simulate the points of impact of the warheads on T. Around these points of impact, we draw circles equal to the lethal radius of the bomb. Then, another stream of pair of random numbers, from an appropriate uniform distribution, are generated representing target points on the target. If these points lie in one of the three circles, they are counted, else rejected. Let M of these lie with in one or more of the lethal circles with their centers at simulated points of impact. Also, let K represents the number of points, out of M, which lie within at least two lethal circles. Then, the expected coverage of the target can be estimated by the sample mean E of

$$E = M/N$$

and the expected overlapping coverage, by the sample mean D of

$$D = K/M$$

Using relevant techniques of statistical inference, as described below, we determine the error in the estimated coverage by Monte Carlo Simulation method.

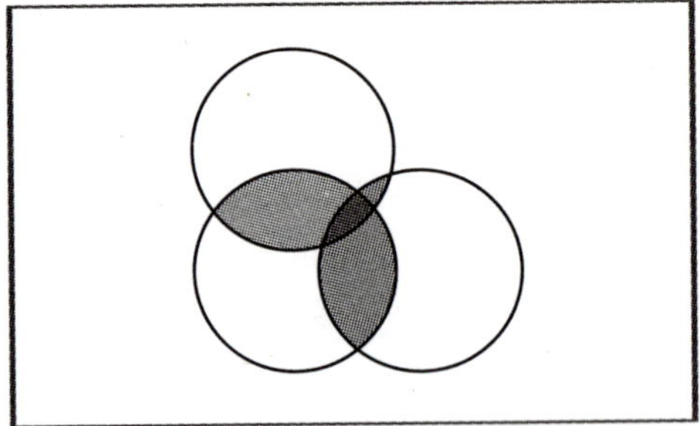

Fig. 4.8: *Overlapping of three bombs on a rectangular target.*

Let j simulation runs be made using different streams of random numbers for simulating the given warheads impact points and N target-points. The sample means $\overline{E}$ may be considered to estimate the expected target coverage and the associated accuracy may be specified by the confidence interval with $100(1-\alpha)\%$ level of confidence. Using technique of statistical inference [79], the end points of this confidence interval turn out to be

$$E \pm t\left(\frac{\alpha}{2}, j-1\right) \cdot \frac{\sigma_E}{\sqrt{j}} \qquad \text{...(4.16)}$$

where $t(\alpha/2, j - 1)$ represents the $100(1 - \alpha)\%$ percentage point of the Student's t-distribution with $(j - 1)$ degrees of freedom; and σ_E^2, the sample variance. Thus, when the (true) expected coverage is estimated by E, the maximum error δ of the estimate with confidence $100\ (1-\alpha)\%$ would be given by (see appendix 4.2)

$$\delta = t(\alpha/2, j-1) \cdot \frac{\sigma_E}{\sqrt{j}} \qquad \text{...(4.17)}$$

It may be noted that j represents the number of observations (simulation runs) and must be distinguished from N which may be regarded as the length of the simulation run. Similarly, the overlapping coverage D may be analyses.

4.5.2 Simulation Model for Missile Attack

In section 1.3.1 of first chapter, we discussed the static computer models of an airfield. Drawing a static model of an airfield is the first step for the study of its denial (damage). In this section we will extend this model, and study the criteria of airfield denial. Monte Carlo technique of simulation is used to find the number of missiles required to be dropped on the runway tracks given in the section 1.3.1, to ascertain a specified level of damage.

Damage to airfield is to such an extent that no aircraft can land or take off from it. It is well known that modern aircraft can land or take off even if a strip of dimensions 15 × 1000m is available. In order to achieve this we choose few aim points on the runway. These aim points are called Desired Mean Point of Impacts (DMPI). These aim points are taken as the centre of the Desired Mean Area of Impacts (DMAIs), which are demarcated on the runway. Let (x_d, y_d) be the co-ordinates of one of the aim points. To find impact point, two normal random numbers x and y are generated (using Box-Mullar method)

$$x = \sqrt{-2\log(u_1)}\ \sin\ (2\pi u_2)$$

$$y = \sqrt{-2\log(u_1)}\ \cos\ (2\pi u_2)$$

where u_1, u_2, are independent uniform random numbers in the interval (0,1). Then co-ordinates of impact point are given by

$$X_1 = x_d + x\sigma_x \qquad \text{...(4.18)}$$

$$Y_1 = y_d + y\sigma_y$$

where σ_x and σ_y are the standard deviation of impact point in x and y directions respectively. In this analysis σ_x and σ_y are the inaccuracies in the distribution of warhead around the aim point. These are called standard deviations along x and y-directions respectively. When σ_x and σ_y are same and equal to σ, we say distribution is circular.

Now let us assume that due to its Circular Error Probable (CEP), warhead aimed at point (x_d, y_d) has fallen on point (x_I, y_I). Let us assume that warhead contains n_b number of bomblets each of lethal radius r_b which after detonation are distributed uniformly within a circle cantered at (x_1, y_1), and of radius R_{wh}, the lethal radius of warhead.

To generate the (x_i, y_i) co-ordinates of the i-th bomblet, we proceed as follows. Take a pair of independent uniform random numbers (v_1, v_2) from different streams of random numbers between 0 and 1 and put,

$$x_1 = (x_1 - R_{wh}) + (2R_{wh})v_1 \text{ and } y_1 = (y_1 - R_{wh}) + (2R_{wh})v_2$$

If this condition is not satisfied, go on generating different pairs of (x_i, y_i) till the condition is satisfied. Condition for bomblet to lie within the lethal circle is given as

$$\sqrt{(x_i - x_1)^2 + (y_i - y_1)^2} \le (R_{wh} - r_b)$$

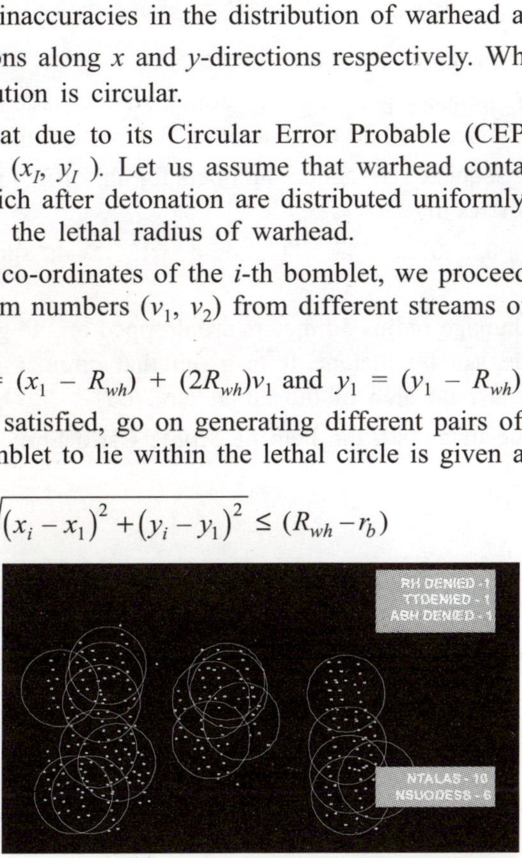

Fig. 4.9: *A computer output of runway denial model using new denial criteria.*

Knowing the position of all the bomblets, it is ascertained that each strip of width W_s has at least one bomblet. If all the DMAIs, are denied, experiment is success otherwise failure. Trial is repeated for large number of times (say 1000 times) and the probability of denial is calculated as the ratio of the number of successes to the number of trials. To ascertain the correct probability of denial programme has been run n times (say 15 times) and the actual probability of denial has been obtained as the average of these n probabilities. Figure 4.9 gives the output of the compute model.

EXERCISE

1. What is discrete simulation? (PTU, 2004)
2. What is a stochastic variable? How does it help in simulation? (PTU, 2004)
3. What is a Monte Carlo techniques? Explain with example. (PTU, 2004)
4. Employ the arithmetic congruential generator, to generate a sequence of 10 random numbers given $r_1 = 971$, $r_2 = 435$ and $m = 1000$.

   (*Hint*). Use $r_{i+2} = (r_i + r_{i+1})$.
5. Describe a procedure to physically generate random numbers on the interval [0,1] with two digit accuracy.
6. Write a computer program for the example 5.
7. Why the random numbers generated by computer are called pseudo random numbers? Discuss the congruence method of generating the random numbers.
8. Test the following sequence of random numbers for uniformity and independence using one of the methods for testing.

   .342, .886, .748,.302, .052, .243, .111, .554, .613, .964, .0033, .465, .777, .732, .406, .165, .767
9. Three bombs of damage radius 50 meters are dropped on the geometric centre of a square target of one side as 100 meters. It is given that circular error probable of bomb is 20 meters. What is total area of the target covered?
10. Compute the value of π with the help of Monte Carlo Simulation method.

APPENDIX 4.1

DERIVATION OF BOX-MULLER METHOD

Probability density function of bi-variate normal distribution for $\mu = 0$ and $\sigma = 1$ is

$$f(x, y) = \frac{1}{2\pi} e^{-(x^2+y^2)/2}.$$

Consider the transformation

$$x = r\cos\theta$$
$$y = r\sin\theta$$

Then

$$p(x, y)dxdy = \frac{1}{2\pi} e^{-\frac{1}{2}r^2} r\,dr\,d\theta$$

Thus r^2, and θ are independently distributed i.e., θ has a uniform distribution in $(0, 2\pi)$ and $r^2/2$ has an exponential distribution. Thus from examples 4.3 and 4.4 , if U_1 and U_2 are uniform random variables from two independent streams we have

$$R = \sqrt{-2\ln(U_1)}$$
$$\theta = 2\pi U_2$$

where R and θ are the random numbers. Hence we get two normal random variables as

$$X = \sqrt{-2\ln(U_1)}\sin(2\pi U_2)$$

$$Y = \sqrt{-2\ln(U_1)}\cos(2\pi U_2)$$

APPENDIX 4.2

SAMPLING DISTRIBUTION OF MEANS

It is generally not possible to find mean of the population which is too large. In this case to estimate the mean, we select few samples of finite size and find mean of each sample and thus total mean of samples. The mean thus obtained may deviates from population mean.

In order to determine the extent to which a sample mean $(\mu_{\bar{x}})$ might differ from the population mean, a parameter $\sigma_{\bar{x}}$ standard error of the mean has to be determined. This is given as

$$\sigma_{\bar{x}} = \sqrt{\frac{\sum(\bar{x} - \bar{\mu}_{\bar{x}})^2}{N}}$$

where N = total number of possible samples. But it is not possible to take total number of possible samples for determining $\sigma_{\bar{x}}$, if population is too large. Thus limited number of samples of finite size are chosen out of the large population, and standard error is estimated as

$$\sigma_{\bar{x}} = \frac{\sigma}{\sqrt{n}} \sqrt{\frac{N-n}{N-1}}$$

where σ = the population standard deviation,

N = population size,

n = sample size.

If population is infinite, standard deviation can be calculated as

$$\sigma_{\bar{x}} = \frac{\sigma}{\sqrt{n}}$$

When the population standard deviation (σ) is known, we can directly compute its standard error of the mean. Thus, the interval estimate may be constructed as

$$\bar{x} - z\sigma_{\bar{x}} < \mu < \bar{x} + z\sigma_{\bar{x}}$$

where $z = \dfrac{x - \mu}{\sigma}$ under the normal curve. For example, for 95% confidence, coefficient $z = 1.96$.

In many situations, not only is the population mean unknown but also the population standard deviation is unknown. For such a case, it appears intuitively that the sample standard deviation (s) is an estimation of the population standard deviation to σ. The computation of s and σ are given by

$$s = \sqrt{\frac{\sum (x-\bar{x})^2}{n}}, \quad \sigma = \sqrt{\frac{\sum (x-\mu)^2}{N}}$$

But although there is a similarity between s and σ, we must remember that one of the important criteria for a statistic to qualify as an estimation is the criterion of unbiasedness. The sample standard deviation is not an unbiased estimator of the population standard deviation.

The unbiased estimator of the population standard deviation is denoted by $\hat{\sigma}$ and is given by

$$\hat{\sigma} = \sqrt{\frac{\sum (x-\bar{x})^2}{n-1}}$$

Thus estimation of the standard error is given by

$$\hat{\sigma}_{\bar{x}} = \frac{\hat{\sigma}}{\sqrt{n}} \qquad \text{(for infinite population)}$$

and

$$\hat{\sigma}_{\bar{x}} = \frac{\hat{\sigma}}{\sqrt{n}}\sqrt{\frac{N-n}{N-1}} \qquad \text{(for finite population)}$$

However,

$$\hat{\sigma}_{\bar{x}} = \frac{\hat{\sigma}}{\sqrt{n}}$$

$$= \frac{s}{\sqrt{n-1}}$$

Thus, if the population standard deviation is unknown, the sampling distribution of means can be assumed to be a approximately normal only when the sample size is relatively large (> 30). In this case interval estimate for large samples is altered slightly and is written as

$$\bar{x} - z\hat{\sigma} \angle \mu \angle \bar{x} + z\hat{\sigma}_{\bar{x}}$$

ESTIMATION USING STUDENT *t*-DISTRIBUTION

When sample size is not large, sampling distribution of means follow for t-distribution in place of normal distribution. t-distribution tends to z-distribution when $n \to \infty$ ($n > 30$). Thus if σ is unknown, and if the sample size is small, the internal estimate **of** the population mean has the following form:

$$\bar{x} - t(\alpha/2, n-1)\sigma_{\bar{x}} \angle \mu \angle \bar{x} + t(\alpha/2, n-1)\sigma_{\bar{x}}$$

Here α is the chance of error and n is degree of freedom in t-distribution (sample size). Like the z-value, the value of t depends on the confidence level. For t-distribution see reference to some standard text book on statistics [54,79].

APPENDIX 4.3

Table 4.6: Area in right tail of a chi-square distribution

Degree of freedom	*0.20*	*0.10*	*0.05*	*0.02*	*0.01*
1	1.642	2.706	3.841	5.412	6.635
2	3.219	4.605	5.991	7.824	9.210
3	4.642	6.251	7.815	9.837	11.345
4	5.989	7.779	9.448	11.668	13.277
5	7.289	9.236	11.070	13.388	15.087
6	8.558	10.645	12.592	15.033	16.812
7	9.803	12.017	14.067	16.622	18.475
8	11.030	13.362	15.507	18.168	20.090
9	12.242	14.684	16.919	19.679	21.666
10	13.442	15.987	18.307	21.161	23.209
11	14.631	17.275	19.675	22.618	24.725
12	15.812	18.549	21.026	24.054	26.217
13	16.985	19.812	22.362	25.472	27.688
14	18.151	21.064	23.685	26.873	29.141
15	19.311	22.307	24.996	28.259	30.578
16	20.465	23.542	26.296	29.633	32.000
17	21.615	24.769	27.587	30.995	33.409
18	22.760	25.989	28.869	32.346	34.805
19	23.900	27.204	30.144	33.687	36.191
20	25.038	28.412	31.410	35.020	37.566
21	26.171	29.615	32.671	36.343	39.932
22	27.301	30.813	33.924	37.659	40.289
23	28.429	32.007	35.172	38.968	41.638
24	29.553	33.196	36.515	40.270	42.980
25	30.675	34.382	37.652	41.566	44.314
26	31.795	35.563	38.885	42.856	45.642
27	32.912	36.741	40.113	44.410	46.963
28	34.027	37.916	41.337	45.419	48.278
29	35.139	39.087	42.557	46.693	49.588
30	36.250	40.256	43.773	47.962	50.892

APPENDIX 4.4

Table 4.7: Kolmogrov-Smirnov critical value

Degree of freedom (N)	$D_{0.10}$	$D_{0.05}$	$D_{0.01}$
1	0.950	0.975	0.995
2	0.776	0.842	0.929
3	0.642	0.708	0.828
4	0.564	0.624	0.733
5	0.510	0.565	0.669
6	0.470	0.521	0.618
7	0.438	0.486	0.577
8	0.411	0.457	0.543
9	0.388	0.432	0.514
10	0.368	0.410	0.490
11	0.352	0.391	0.468
12	0.338	0.375	0.450
13	0.325	0.361	0.433
14	0.314	0.349	0.418
15	0.304	0.338	0.404
16	0.295	0.328	0.392
17	0.286	0.318	0.381
18	0.278	0.309	0.371
19	0.272	0.301	0.363
20	0.264	0.294	0.356
25	0.240	0.270	0.320
30	0.220	0.240	0.290
35	0.210	0.230	0.270
Over 35	$\frac{1.22}{\sqrt{N}}$	$\frac{1.36}{\sqrt{N}}$	$\frac{1.63}{\sqrt{N}}$

APPENDIX 4.5

EMPIRICAL DISTRIBUTION FUNCTION

In statistics, an **empirical distribution function** is a cumulative probability distribution function that concentrates probability $1/n$ at each of the n numbers in a sample.

Let $X_1, \ldots, X_n$ be random variables with realizations $x_i \in R$, $i = 1, \ldots, n \in N$.

The empirical distribution function $F_n(x)$ based on sample $x_1, \ldots, x_n$ is a step function defined by

$$F_n(x) = \frac{\text{number of elements in the sample} \leq x}{n} = \sum_{i=1}^{n} I_A(x_i \leq x),$$

where I_A is an indicator function.

INDICATOR FUNCTION

In *mathematics*, an **indicator function** or a **characteristic function** is a function defined on a *set* X that indicates membership of an element in a *subset* A of X.

The indicator function of a subset A of a set X is a function

$$I_A : X \to \{0,1\}$$

defined as

$$I_A(x) = \begin{cases} 1 \text{ if } x \in A, \\ 0 \text{ if } x \notin A. \end{cases}$$

❑❑❑

5

CONTINUOUS SYSTEM SIMULATION

So far we have discussed problems and techniques which are stochastic in nature. In the present chapter, problems of continuous nature will be studied. Modeling and simulation of various problems like pursuit evasion of aircraft, fluid flow, flight dynamics and so on come under continuous system simulation. Continuous system generally varies with time and is dynamic. As it has been mentioned earlier too, there is no fixed method for modeling a particular problem. However, one of the basic tools for modeling beyond doubt is mathematics. One can model any situation with the help of mathematics. Sometime mathematical models cannot be solved analytically, as we have seen in the case of hanging wheel of a vehicle. There was a time when approximations were made to simplify the mathematical equation of the problem. But now with the advances made in modern computers, any continuous model can be converted to digitization and worked out with the help of computer programming. We had assumed in this book that reader is conversant with computer programming with special reference to C++ language.

In first few sections, we will discuss the basic mathematical tools to be used in continuous simulation. It is not possible to study these tools in details, as full book is needed for each topic, but attempts will be made to cover the topic briefly and reference to required book will be made wherever required.

5.1 What is Continuous Simulation ?

In chapter two and four we have studied discrete modeling and simulation. One of the definitions of a discrete variable is that it contains at most finitely many of the values, in a finite interval on a real number line. Thus there can be certain portions on the number line where no value of discrete variable lies. But in the case of continuous variable, it has infinite number of values in a finite interval. By continuous we mean uninterrupted, remaining together, not broken or smooth flowing. As per the definition given above there is no interval, how so ever small, on number line where a value of continuous variable is not there. If $f(x)$ is a continuous function, then $y = f(x)$ is a smooth curve in x-y plane. We give below the definition of continuity as is taught to us at school level.

Definition of Continuity: *A function f(x) is said to be continuous at* $x = x_0$ *if* $\lim_{x \to x_0} f(x) = f(x_0)$; *equivalently, given any* $\varepsilon > 0$ *there exists a* $\delta > 0$ *such that* $|f(x) - f(x_0)| < \varepsilon$ *wherever* $|x - x_0| < \delta$.

We are not going to discuss types of continuity as it is not in the purview of the present book. In the universe, there are number of examples of continuity, such as fluid flow, projectile motion, aircraft flight, and so on. Any medium which looks continuous, when looked at macroscopically, becomes discrete when we see it microscopically. For example air in a room is continuous but when we see it with electron microscope, we find it is nothing but consisting of molecules of different gases moving randomly, to and fro. Aim of this chapter is how to simulate continuous problems and various tools required for the modeling of continuous systems. One of the basic tools required for the analysis of continuous systems is Mathematics. A good system analyst has to be very strong in Mathematics. Along with mathematics, numerical techniques and computer programming is also essential. Although we assume that students of this book are conversant with these tools, yet attempt will be made to explain details of models wherever it is required.

5.2 Modeling of Fluid Flow

Most common example of continuous state is fluid flow. In this universe, most of the materials are in fluid state. Literal meaning of fluid is 'which flows'. Flow of water in rivers and air in atmosphere are examples of fluid flow. But is there any law which fluid flow has to obey? Whole of this universe is bound by some laws. Earth revolves around sun as per some Mathematical law and sun moves towards some galaxy as per some other law. In order to model these phenomena, we have to know what laws behind their motion are? This is called physics of the problem, which one has to know before one takes up the modeling of a system. That means, before making the Mathematical model of a scenario, we have to know its physics. By physics we mean its total working. Even in earlier chapters, it has been mentioned that in order to model any system, first we have to fully understand the system. Let us first understand the physics of fluid flow.

It is well known that fluid flow obeys three basic laws, called conservation laws that is, conservation of mass, momentum and energy. These laws are based on three basic principles of Physics. Conservation of mass says, it can neither be created nor destroyed. Second law is after famous scientist Newton, and is called second law of motion. This law is defined as, rate of change of momentum of a system is equal to the applied force on it and is in the same direction in which force is applied. Third law says, energy of system remains same, only it changes from one form to other form. Equations governing these laws are called continuity, momentum and energy equations. In the next section we will make mathematical models of these laws of nature.

5.2.1 Equation of Continuity of Fluids

In order to model equation of continuity, we assume an infinitesimal cubical volume in fluid in motion. Continuity equation states that in an enclosed fluid volume, amount of fluid entering in it is same as fluid leaving, until and unless there is source or sink in the volume. By source or sink we mean, neither fluid is created nor destroyed in the volume. Let volume of the cube in the fluid be V, and surface area be S. Then the changes in the mass of fluid contained in this cube is equal to the net quantity of fluid flowing in and out from its boundary surfaces. We assume that this cube is very small of infinitesimal size, whose sides are of length dx, dy and dz and volume $V = dx \cdot dy \cdot dz$. Then change in the mass of fluid in the volume V in time dt, resulting from change in its density ρ, is,

$$dm = d\,(\rho \cdot dx \cdot dy \cdot dz)$$

This can be written as,

$$dm = \frac{\partial}{\partial t}(\rho dx \cdot dy \cdot dz) \cdot dt$$

$$= \frac{\partial \rho}{\partial t} dx \cdot dy \cdot dz \cdot dt \qquad ...(5.1)$$

since x, y, z, and t are independent coordinates in three-dimensional space.

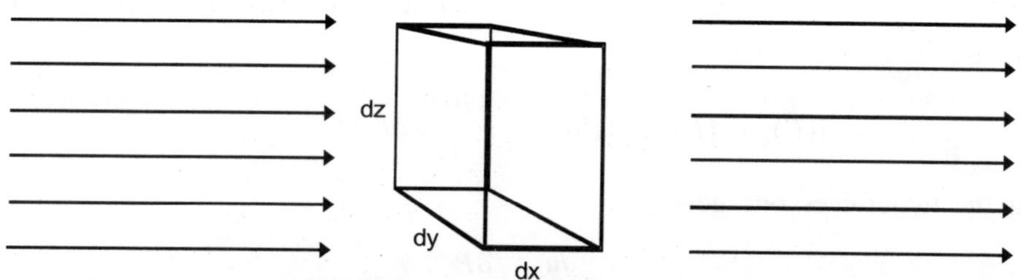

Fig. 5.1: *A small cube in the fluid.*

Change in the mass of fluid in the cube is due to the movement of fluid in and out of its surfaces. Let velocities of the fluid in x, y, z directions be u, v and w respectively, where these velocity components are function of space coordinators x, y, z and time t. If in time dt, mass $(\rho u)_x$ enters from the surface $dy \cdot dz$ and $(\rho u)_{x+dx}$ leaves from the other parallel face then,

$$[(\rho u)_x - (\rho u)_{x+dx}]\, dt\, dy\, dz = -\frac{\partial}{\partial x}(\rho u)\, dt\, dx\, dy\, dz \qquad ...(5.2)$$

where higher order terms in the expansion of $(\rho u)_{x+dx}$ have been neglected, being comparatively small.

Similarly we get two more relations for movement of fluid along y- and z-directions i.e.,

$$[(\rho v)_y - (\rho v)_{y+dy}]\, dt\, dx dz = -\frac{\partial}{\partial y}(\rho v)\, dt\, dx\, dy\, dz$$

$$[(\rho w)_z - (\rho w)_{z+dz}]\, dt\, dx\, dy = -\frac{\partial}{\partial z}(\rho w)\, dt\, dx\, dy\, dz \qquad ...(5.3)$$

Sum of the fluid motion out of three surfaces of the cube is equal to rate of change of mass in side the cube. Thus from equations (5.1), (5.2) and (5.3) one gets,

$$\frac{\partial \rho}{\partial t} + \frac{\partial}{\partial x}(\rho u) + \frac{\partial}{\partial y}(\rho v) + \frac{\partial}{\partial z}(\rho w) = 0 \qquad ...(5.4)$$

This equation is the equation of continuity. This result in vector form can be written as,

$$\frac{\partial \rho}{\partial t} + \text{div}(\rho \vec{v}) = 0$$

where the velocity $\vec{v}$ has the components u, v, w in cartesian coordinates and *div* is the divergence.

5.2.2 Equation of Momentum of Fluids

Conservation of momentum is based on Newton's second law of motion, or conservation of momentum. It is convenient to consider the forces acting on the element of volume *dxdydz* moving with the fluid rather than an element fixed in space. Considering x-component of momentum, the acceleration of the element is given by the total derivative of velocity of fluid in x-direction i.e., u which is $\frac{du}{dt}$. Product of this acceleration with mass ρ *dxdydz* of the element of the moving fluid, according to second law of Newton, must be equal to force acting on the volume in the x-direction. This force is nothing but difference of pressure P, acting on the two parallel faces of area *dydz*. This pressure gradient is given as

$$[(P)_x - (P)_{x+dx}]\, dydz = -\frac{\partial P}{\partial x} dxdydz$$

Equating the two forces one gets

$$\rho\frac{du}{dt} = \frac{\partial P}{\partial x}$$

Expressing $\frac{du}{dt}$ in terms of partial derivatives, one gets

$$\rho\frac{\partial u}{\partial t} + \rho u\frac{\partial u}{\partial x} + \rho v\frac{\partial u}{\partial y} + \rho w\frac{\partial u}{\partial z} = -\frac{\partial P}{\partial x} \qquad ...(5.5a)$$

Similarly two equations in y- and z-directions are obtained as,

$$\rho\frac{\partial v}{\partial t} + \rho u\frac{\partial v}{\partial x} + \rho v\frac{\partial v}{\partial y} + \rho w\frac{\partial v}{\partial z} = -\frac{\partial P}{\partial y} \qquad ...(5.5b)$$

$$\rho\frac{\partial w}{\partial t} + \rho u\frac{\partial w}{\partial x} + \rho v\frac{\partial w}{\partial y} + \rho w\frac{\partial w}{\partial z} = -\frac{\partial P}{\partial z} \qquad ...(5.5c)$$

These equations when written in vector form are,

$$\rho\frac{d\vec{v}}{dt} = \rho\frac{\partial \vec{v}}{\partial t} + \rho(\vec{v}\cdot\text{grad})\vec{v} = -\text{grad}\, P \qquad ...(5.6)$$

5.2.3 Equation of Energy of Fluids

Following the method used in deriving of momentum equation, we consider the element volume moving with the fluid and enclosing the fixed mass of fluid. Total energy per unit mass of the fluid consists of kinetic energy plus internal energy E, which is the sum of thermal and chemical energies. The change in energy in time dt for the element of volume *dxdydz* is

$$\rho\frac{d}{dt}\left[E + \frac{1}{2}(u^2 + v^2 + w^2)\right] dt\, dx\, dy\, dz$$

Where the total time derivative is used to account for the displacement of the element during the interval. This change in energy must be equal to the work done by the fluid on the surfaces of element volume. The work done in time dt on an area $dydz$ in motion along x-direction, is the product of force and displacement, or $Pu\ dtdydz$, and net amount of work done on the two faces of the volume is

$$[(Pu)_x - (Pu)_{x+dx}]\,dtdydz = -\frac{\partial(Pu)}{\partial x}\,dt\,dx\,dy\,dz$$

The work done on other faces is obtained in the same way, and equating the total work on all the faces to the change in energy, one gets

$$\rho\frac{d}{dt}\left[E+\frac{1}{2}(u^2+v^2+w^2)\right] = -\left[\frac{\partial}{\partial x}(Pu)+\frac{\partial}{\partial y}(Pv)+\frac{\partial}{\partial z}(Pw)\right] \qquad ...(5.7)$$

which is the equation for energy conservation. This equation in vector notation is given as,

$$\rho\frac{d}{dt}\left[E+\frac{1}{2}(\vec{v}\cdot\vec{v})\right] = -\mathrm{div}(P\vec{v}) \qquad ...(5.7a)$$

where $\vec{v} = (u, v, w)$.

The equations (5.4),(5.6) and (5.7) are non-linear equations and can not be solved by analytic methods. Therefore, one has to go for numerical techniques. Some of the techniques for numerical computations and their programs in C++ language are given in appendix 5.1 for the convenience of reader. Those who want to study this subject in details can refer to some book on computational fluid dynamics.

5.3 Dynamic Model of Hanging Car Wheel

In chapter one, we discussed the case of a hanging wheel of a vehicle. Using Newton's second law of motions, physical model has been expressed in terms of mathematical equations as,

$$M\frac{d^2x}{dt^2} + D\frac{dx}{dt} + Kx = KF(t) \qquad ...(5.8)$$

where x = the distance moved,

M = mass of the wheel,

K = stiffness of the spring,

D = damping force of the shock absorber.

Dividing equation (5.8) by M, one gets,

$$\frac{d^2x}{dt^2} + \frac{D}{M}\frac{dx}{dt} + \frac{K}{M}x = \frac{K}{M}F(t)$$

which can be written as

$$\frac{d^2x}{dt^2} + 2\zeta\omega\frac{dx}{dt} + \omega^2 x = \omega^2 F(t) \qquad ...(5.8a)$$

where $2\zeta\omega = D/M$, $\omega^2 = K/M$

We integrate this equation using Runge-Kutta method of fourth order (see appendix 5.1). In order to use this method, equation (5.8*a*) is to be converted in first order equations. Equation (5.8*a*) is converted to two first order differential equation as follows,

$$\frac{dx}{dt} = y$$

$$\frac{dy}{dt} = f(x, y, t) = \omega^2 F(t) - 2\zeta\omega y - \omega^2 x \qquad \text{...(5.8}b\text{)}$$

These are two homogeneous first order differential equations in x and y. Runge-Kutta method gives us

$$m_1 = y_i$$

$$m_2 = y_i + \frac{m_1 h}{2}$$

$$m_3 = y_i + \frac{m_2 h}{2}$$

$$m_4 = y_i + m_3 h$$

$$n_1 = f(x_i,\ y_i)$$

$$n_2 = f(x_i + \frac{n_1 h}{2},\ y_i + \frac{m_1 h}{2})$$

$$n_3 = f(x_i + \frac{n_2 h}{2},\ y_i + \frac{m_2 h}{2})$$

$$n_4 = f(x_i + n_3 h,\ y_i + m_3 h)$$

$$x_{i+1} = x_i + \left(\frac{m_1 + 2m_2 + 2m_3 + m_4}{6}\right) h$$

$$y_{i+1} = y_i + \left(\frac{n_1 + 2n_2 + 2n_3 + n_4}{6}\right) h \qquad \text{...(5.9)}$$

$$h = \Delta t$$

Using the above algorithm, and assigning some values to $F(t)$, D, and K one can compute the results of equation (5.8*b*). Figure 5.2 gives the variation of x verses ωt, when a steady force F is applied to wheel at time $t = 0$.

Figure 5.2 shows how x varies in response to a steady force applied at time $t = 0$. Solutions are shown for different values of ζ. It can be seen from the curves for different values of ζ that oscillations of the wheel increase as ζ decreases. For oscillations to be minimum, $\zeta \geq 1$, that means $D^2 \geq 4MK$. Here ζ is called the damping force. When motion is oscillatory, the frequency of oscillation is determined from the formula

$$\omega = 2\pi f$$

where f is the number of cycles per second.

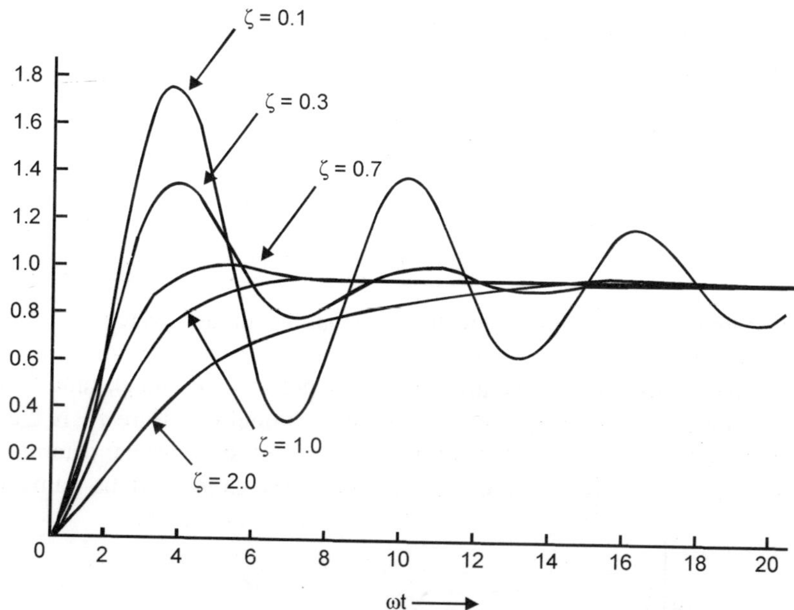

Fig. 5.2: *Graph showing displacement vs. time. (Courtesy Geoffery Gordon, System Simulation)*

5.4 Modeling of Shock Waves

If one is living near an airfield where supersonic aircraft are flying, banging sounds that shatters the window pains of one's house is a frequent phenomenon. What are these ear breaking sounds? These are called sonic booms and are created whenever a supersonic aircraft crosses limits of subsonic speed to a supersonic speed. This speed barrier is called *sonic barrier.* At this point, where aircraft changes from sonic to supersonic speed, shock waves are created. It is well known that shock waves are present when ever there is a supersonic motion, or there is a blast. Problem of this section is to study, how to model the formation of shock waves. First of all, it is important to understand, what these shock waves are after all?

All the weapons contain explosives as a basic ingredient. Explosives generate high pressure waves called shock waves, which cause damage to the targets. In this section, we will give briefly the theory of shock waves. We know that due to detonation of explosives, buildings and other structures around get damaged. This damage occurs due to impact of the shock waves with the targets (buildings and structures). Almost similar type of effects are felt, when a supersonic aircraft flies over our heads. After all what are these shock waves?

Generation of large amount of energy in a small volume and short time duration produces shock waves. As an example, detonation of an explosive generates large amount of energy and produces a shock in the atmosphere. Shock waves in air produced by the detonation of explosives are given a different naming i.e., Blast waves. Shock waves can also be generated by means, other than explosions, for example spark, mechanical impact, bursting of boilers etc. Shock waves are also called pressure wave or compression wave as pressure and density of the medium is much more behind the shock front than that ahead of it. In Fig. 5.3, a shock profile has been shown. This curve shows variation of pressure behind shock as a function of time. It is a replica of an oscilloscope record connected with the pressure gauge, used for recording the blast pressure.

Table 5.1: Lethal radii computed from the Scaling Law for various types of damage

Targets and types of damage	*Persons (kill)*	*Persons (ear drum burst)*	*Soft skinned vehicles (damage)*	*Window glass (breaking)*	*Antitank mines*	*Antipersonnel mine*
Pressure required (kg/cm²)	4.5332	2.0332	1.3132	1.1332	7.6632	1.6632

One of the most important property of shock waves is, the sudden increase in pressure behind it's front. This pressure (called peak over pressure) is mainly responsible for damage to structures. Some of the important parameters of shock wave are: peak over pressure ($P_1 - P_0$), time duration of positive pressure τ, time duration of negative pressure τ' and shock impulse I. Here P_1 is the pressure behind the shock front and P_0 is ahead of it. These parameters are responsible for the determination of damage due to blast. In Table 5.1, we give pressure range which can cause various damages.

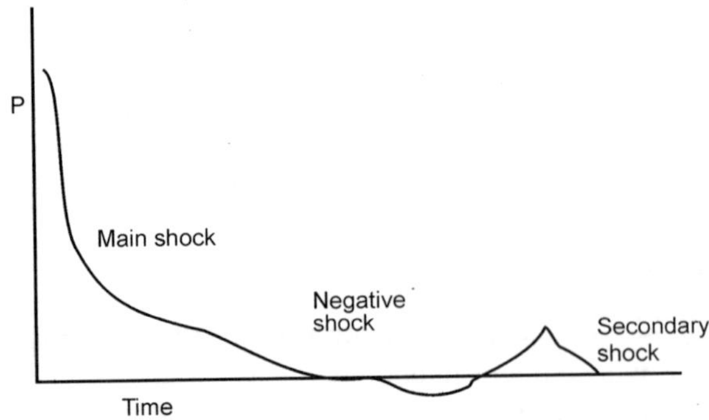

Fig. 5.3: *Shock pressure record taken from oscilloscope.*

Peak over pressure ($P_1 - P_0$), is the pressure across the shock front and is denoted by P in Fig. 5.3 and time duration τ is the time for which this pressure remains positive. Impulse I, of the shock is given as

$$I = \int_0^{\tau} (P_1 - P_0)\, dt \qquad ...(5.10)$$

where $(P_1 - P_0)$, is a function of time t and is of the form $P_1 = P_m e^{-at}$. Impulse is an important parameter for the damage.

5.4.1 Shock Waves Produced by Supersonic Motion of a Body

If a projectile or an aircraft moves in air with the velocity greater than the speed of sound, a shock is produced at its head. It may be attached or detached depending on the shape of the projectile[35, 45] As it is clear from the names, attached shock touches the nose of the projectile whereas detached shock moves a bit ahead of it. How this shock is produced is described below.

It is known that any disturbance produced in air moves with the speed of sound in all the directions Any aircraft, which is moving in air with a velocity v, produces sound waves, which move with the velocity $\boldsymbol{a}$, outward in the form of a spherical envelope. If the speed of the aircraft is less than or equal to that of speed of sound then sound wave fronts will be as given in Fig. 5.4a (here $\boldsymbol{v} = \boldsymbol{a}$) But if it moves with the speed greater than sound, the situation will be as in Fig. 5.4b. In this figure three circles in 5.4*a* and 5.4*b* are the positions of sound waves starting respectively from points *A*, *B* and *C* at time when aircraft reaches point *O*. For details see Singh (1988). If we draw a tangent to all these circles, it will pass through point *O*. Thus envelope of these circles is a cone with half angle α. In Fig. 5.5, it can be seen that aircraft moves with speed ***v*** which is greater than sound speed ***a***, thus all the disturbance due to the motion of the aircraft remains confined in a cone whose semi-vertex angle α is given by,

$$\alpha = \sin^{-1}(1/M)$$

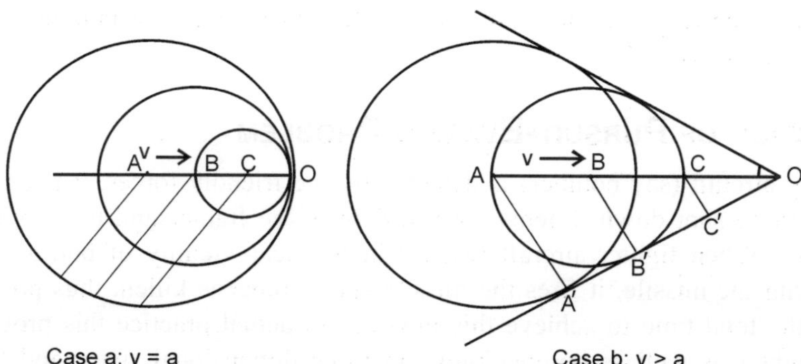

Fig. 5.4: *Disturbance due to a body moving with sonic and supersonic velocity.*

where ***M= v/a.*** Ratio ***M*** is called Mach number. Derivation of this equation is clear from Fig. 5.4*b*. By the time aircraft moves from point *A* (*B* or *C*) to *O*, sound wave produced at *A* (*B* or *C*) reaches the point *A′* (*B′* or *C′*), where line *OA′* is tangent to all the three circles with centers at *A,B* and *C* respectively. Thus,

$$\sin\alpha = \frac{AA'}{OA'} = \frac{\vec{a}t}{\vec{v}t} = \frac{1}{M}$$

Fig. 5.5: *Aircraft disturbances cannot be heard outside the Mach cone. (Courtesy G. Gamow; frontiers of Physics)*

Thus we see that disturbance in case of supcrsonic motion remain confined in side the Mach cone with semi-vertex angle α (see Fig. 5.5). It is known that when a supersonic aircraft is seen by a person on the ground, its sound is not heard, but when it has crossed the person's head, suddenly a roaring sound is heard. This is because earlier, person was out of the range of the Mach cone and heard no noise but as soon as aircraft crossed over his head, he came inside the Mach cone, in which whole disturbance is confined and heard the roaring sound. Surface of the Mach cone is nothing but the shock front. As we have stated earlier, the pressure behind the shock front is very high, by which even buildings are damaged. This shock wave is weaker than the blast wave (that produced by explosives), that is the reason it has been given different name after a famous scientist Ernest Mach and is called Mach wave.

5.5 Simulation of Pursuit-Evasion Problem

When an enemy aircraft (say bomber) is detected by the friendly forces, it is chased by the friendly fighter aircraft to be shot down. Enemy aircraft detects the fighter and tries to avoid it by adopting different tactics. When fighter aircraft reaches in the near vicinity of bomber, which is a critical distance for firing the missile, it fires the missile and bomber is killed, thus problem ends. Problem is to compute the total time to achieve this mission. In actual practice this problem is too complex to model, as fighter as well as bomber move in three dimensional space and there moves are not known. In order to make the problem simple, we assume that bomber and fighter, both fly in the same horizontal plane and flight path of bomber is a known parameter.

Let us take a two dimensional plane as *xy*-plane, for the combat and initial position of fighter and bomber respectively is $y_f = 50$ and $x_b = 100$, where subscript *f* and *b* respectively indicate values for fighter and bomber. It is assumed that flight path of bomber makes an angle θ° with *x*-axis. After each interval of time (say a second), fighter detects the current position of bomber, and computes its future position and moves towards that position. Equations of flight path of bomber can be written as,

$$x_b(t) = 100 + v_b \cos(\theta) \cdot t$$

$$y_b(t) = v_b \sin(\theta) \cdot t$$

if x_f, y_f are coordinates of fighter at time *t*, then the distance between fighter and bomber at time *t* is,

$$\text{distance} = \sqrt{(y_b(t) - y_f(t))^2 + (x_b(t) - x_f(t))^2}$$

If ϕ be the angle which fighter to bomber path at time *t*, makes with *x*-axis then (Fig. 5.6),

$$\frac{dx_f}{dt} = v_f \cos(\phi) \cdot t$$

$$\frac{dy_f}{dt} = v_f \sin(\phi) \cdot t$$

In order to solve these equations for (x_f, y_f), we convert these equations into difference equations, thus,

$$x_f(t+1) = x_f(t) + v_f \cos(\phi)$$

$$y_f(t+1) = y_f(t) + v_f \sin(\phi)$$

where

$$\tan(\phi) = \frac{y_b(t) - y_f(t)}{x_b(t) - x_f(t)}$$

This problem although simple can not be solved analytically. We adopt the technique of numerical computation. Conditions are, when distance (dist) ≤ 10km, missile is fired and bomber is destroyed. If this distance is not achieved in 12 minutes from the time of detection, bomber escapes. We have assumed that velocity of bomber is 20 km/m and that of fighter is 25 km/m. It is observed that target is killed in six minutes with given velocities. Results of computation are given in Fig. 5.6 and a computer program is given as program 5.1.

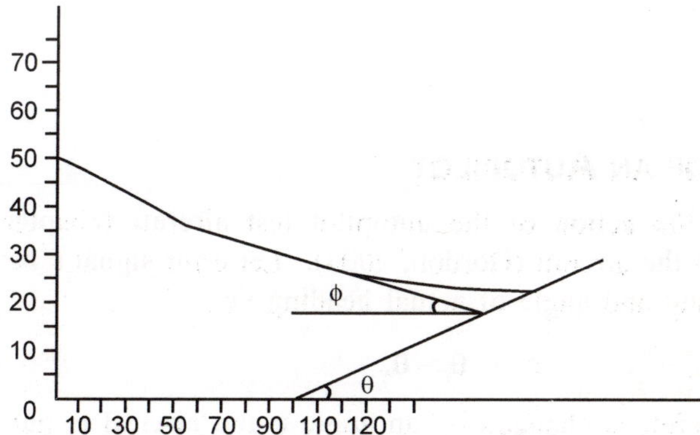

Fig. 5.6: *Pursuit-Evasion problem.*

Program 5.1: Pursuit-Evasion Problem of Two Aircraft

```
#include <iostream.h>
#include <math.h>
#include <conio.h>
void main(void)
{
    float xb=100.00, yb=0, xf = 0,yf=50.0,dist,psi=0.2,theta,vb=20.0,
      vf=25.0, tlim = 12.0;
    int t;
    theta = atan(0.5);
  for (t=0; t<=16; t++)
    {
    xb=xb +vb*t*cos(psi);
    yb=yb+vb*t*sin(psi);
    xf = xf + vf*t*cos(theta);
    yf = yf -t*vf*sin(theta);
    dist=sqrt((yb-yf)*(yb-yf)+(xb-xf)*(xb-xf));
```

```
        theta = atan((yf-yb)/(xb-xf));
        cout<<"t=" << t <<"\n";
        cout<<"xb="<<xb<<"yb="<<yb<<"xf="<<xf<<"yf="<<yf<<"\n";
        cout<< '\t'<<"dist="<< dist <<'\t'<<"theta=" <<theta<<endl;
        cout<<"\n";
        if ( t> tlim) {cout << "target escapes, game over";break;}
         else if(dist<=10.00)
         {cout<<"target killed\n";break;}
          else
        continue;
        }
}
```

5.6 Simulation of an Autopilot

In order to simulate the action of the autopilot test aircraft (chapter zero), we first construct mathematical model of the aircraft (Gordon, 2004). Let error signal ε be the difference between the angle of desired heading and angle of actual heading i.e.,

$$\varepsilon = \theta_i - \theta_0 \qquad \text{...(5.11)}$$

Rudder of the aircraft is changed by an angle ε and aircraft is put on the right track. Due to change of this angle a torque force is applied to the airframe which turns its direction. Just like the case of hanging wheel of automobile, this torque τ, is resisted by a force called viscous drag, which is proportional to the angular velocity of the aircraft. The torque acting on the aircraft can be represented as

$$\tau = K\varepsilon - D\dot{\theta}_0 \qquad \text{...(5.12)}$$

where K and D are constants. First term on the right is the torque produced by the rudder and second is the viscous drag. Now angular momentum of the aircraft is proportional to applied torque. Proportionality constant in this case is the inertia of the aircraft, say I. Thus we have,

$$I\ddot{\theta}_0 + D\dot{\theta}_0 + K\theta_0 = K\theta_i \qquad \text{...(5.13)}$$

Dividing this equation by I on both sides and making the following substitutions,

$$2\zeta\omega = \frac{D}{I},\ \omega^2 = \frac{K}{I}, \qquad \text{...(5.14)}$$

Thus equation (5.13) reduces to,

$$\ddot{\theta}_0 + 2\zeta\omega\dot{\theta}_0 + \omega^2\theta_0 = \omega^2\theta_i \qquad \text{...(5.15)}$$

This is the second order differential equation and is same as equation (5.8*a*), and its solution is provided in section (5.3). The result shows that the aircraft will have oscillatory motion for $\zeta \le 1$ which implies,

$$D^2 \le 4KI$$

There can be hundreds of example in continuous simulation and it is not possible to discuss all of them. We have studied some of these, just to illustrate, what is continuous simulation.

5.7 MODELING OF PROJECTILE TRAJECTORY

Computation of trajectory of a projectile is often required while modeling various problems related with weapon modeling. In this section, we give a mathematical model for computing the trajectory of a projectile.

The following assumptions have been made for the computation of the trajectories.

(*a*) The aerodynamic force acting on the projectile is the drag force (which includes various forces due to parachutes) acting opposite to the direction of the velocity vector.

(*b*) Indian standard atmosphere, sea level condition, is assumed.

The origin of reference frame for the computation of the trajectories is considered to be positioned at point of ejection of the projectile. It's Y-axis is vertically downwards and the X is along the horizontal direction.

A two dimensional point mass trajectory model has been used for the computation of the flight paths of the projectile, and the equations of which are as given below:

$$m\frac{d^2x}{dt^2} = -\frac{1}{2}\rho SC_D V^2 \cos(\theta) \qquad ...(5.16)$$

$$m\frac{d^2y}{dt^2} = -\frac{1}{2}\rho SC_D V^2 \sin(\theta) + mg \qquad ...(5.17)$$

$$\theta = \tan^{-1}(dy/dx) \qquad ...(5.18)$$

where

θ = angle of elevation,

C_D = drag coefficient,

ρ = density of air,

g = acceleration due to gravity,

m = mass of the body.

It is known that drag coefficient is a function of the velocity of the projectile.

EXERCISE

1. What is continuous simulation? (PTU, 2004)
2. Reproduce the automobile suspension problem with the assumption that the damping force of the shock absorber is equal to $5.6(\dot{x}-0.05\ddot{x})$.
3. Write the steps for integration of following differential equations by Runge-Kutta method.

$$\frac{dx}{dt} = f(x, y, t)$$

$$\frac{dy}{dt} = g(x, y, t)$$

APPENDIX 5.1

A.1 Difference Equations and Numerical Methods

In the above section, we have modeled the fluid flow which is in the form of partial differential equations. In order to find numerical solution of these equations one has to convert these equations in the form of difference equations. For finding the numerical solution of differential equations, one has to convert differential equations to difference equations. Let us take a simple example, where we convert a first order differential equation to a difference equation. Let the equation be

$$\frac{dy}{dx} = ax + by + c \qquad ...(A.1)$$

In order to convert it into a difference equation, we construct a rectangular mesh, so that each rectangle of mesh is of size $dx \cdot dy$ (Fig. A.1).

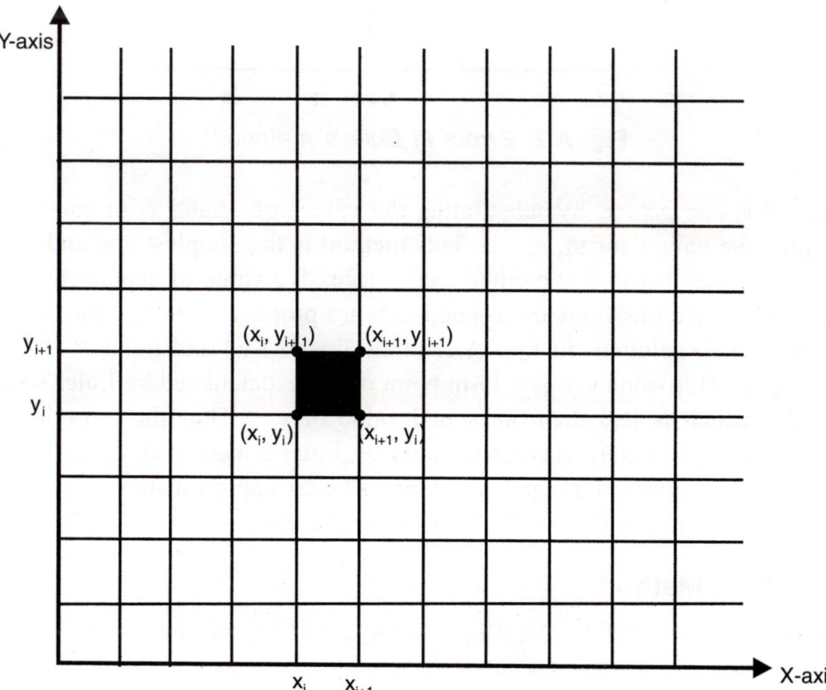

Fig. A.1: *Grid formation in X-Y plane.*

In the Fig. A.1, point x_{i+1} is nothing but $x + \Delta x$, and x_i, y_{i+1} is $x_i, y_i + \Delta y$. Using this nomenclature, we can express differential equation (A.1) as

$$\frac{dy}{dx} = \frac{\Delta y}{\Delta x} = \frac{y_{i+1} - y_i}{x_{i+1} - x_i}$$

Thus differential equation (A.1) becomes at i-th grid as

$$y_{i+1} = y_i + \Delta x(ax_i + by_i + c)$$

If initial values are given as when $x = x_0$, $y = y_0$, then

$$y_1 = y_0 + \Delta x(ax_0 + by_0 + c)$$

$$y_2 = y_1 + \Delta x(ax_1 + by_1 + c)$$

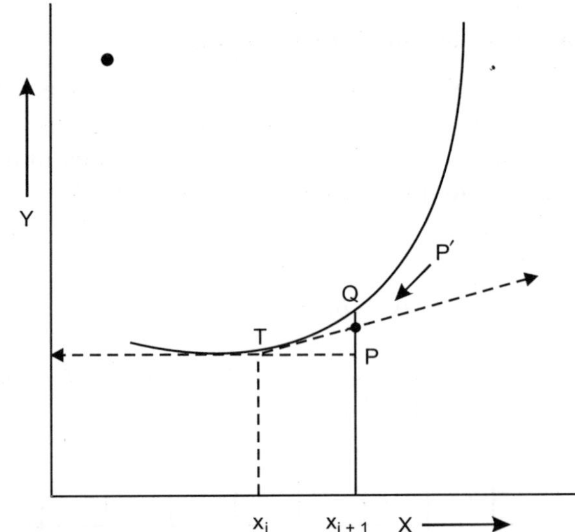

Fig. A.2: *Errors in Euler's method.*

Now $x_1 = x_0 + \Delta x$, one gets y_2 by substituting the values of y_1 and x_1 in equation for y_2. Same way we can compute the values for y_3, y_4,… . This method is the simplest one and is known *Euler's method*. We can easily see that in this method each successive value of dependent variable depends on the previous value. Hence inaccuracies cropped in y_i are propagated to y_{i+1} and slowly and slowly, it deviates from the actual solution. In Fig. A.2, point T is (x_i, y_i) and point P is (x_{i+1}, y_i) where as point Q is (x_{i+1}, y_{i+1}). But value y at $(i + 1)$-th point of grid, calculated by Euler's method (tangent at point T) is at P' which is less than the actual value of y_{i+1}. Thus the error in y_{i+1} that is QP' is added to each step and ultimately curve computed by Euler's method deviates from actual curve. In section A.3 we give an improved method which gives better approximation as compared to Euler's method. A computer program of Euler's method is given below.

Program A.1: Euler's Method

```
/* Computer program to integrate an ordinary differential equation by Euler's
method*/
    #include <iostream.h>
    main()
```

```
{
 int i,n;
float x,y,xp,h,dy;
float func(float, float);
cout<<"\nSolution by Euler's Method\n\n";
/* Reading Initial data*/
cout<<"\n Input Initial values of x and y\n";
cin>>x>>y;
cout<<"input x for which y is required\n";
cin>>xp;
cout<<"Input step size h\n";
cin>>h;
/*Compute number of steps required*/
n=(int)((xp-x)/h+0.5);
/*Compute y recursively at each step*/
for(i=1;i<=n;i++)
{dy=h*func(x,y);
x=x+h;
y=y+dy;
cout<<i<,x<<y<<"\n";
}
/*Write the final results*/
cout<<"Value of y at x="<<x<<"is"<<y<<"\n";
}//End of main()
float func(float x, float y)
{
float f;
f =2.0 *y/x;
return(f)
}
```

All numerical techniques for solving differential equations of the type $\frac{d^n y}{dx_n} = f(x, y)$ involve a series of estimation of $y(x)$ starting from the given conditions. There are two basic approaches that could be used to estimate the value of $y(x)$. They are known as *one-step methods* and *multiple step methods*.

In one step methods, we use the information from only one preceding point i.e., to estimate the value y_i, we need the condition at the previous point y_{i-1} only. Multi step methods use information at two or more previous steps to estimate a value. In this section we will discuss some of the integration techniques for ordinary differential equations.

A.2 Taylor Series Method

According to Taylor series, we can expand a function $y(x)$ about a point $x = x_0$ as

$$y(x) = y(x_0) + (x - x_0)y'(x_0) + (x - x_0)^2 \frac{y''(x_0)}{2!} + \cdots + (x - x_0)n\frac{y^n(x_0)}{n!} \tag{A.2}$$

where $y^n(x_0)$ is the n-th derivative of $y(x)$, evaluated at $x = x_0$. The value of $y(x)$ is known if we know its derivatives. To understand the method let us consider a differential equation,

$$\frac{dy}{dx} = y' = 2x^2 + 3y^3 + 5 \tag{A.3}$$

with the initial conditions $y(0) = 0.5$ at $x = 0$. Now

$$y'' = 4x + 9y^2 y'$$

$$y''' = 4 + 18y(y')^2 + 9y^2 y''$$

At $x = 0$, $y(0) = 0.5$ and therefore

$$y'(0) = 5.625,$$
$$y''(0) = 9\ (1)\ (8) = 12.656,$$
$$y'''(0) = 4 + 18(0.5)\ (5.625)^2 + 9.(0.25)\ (12.656)$$
$$= 4 + 284.765 + 28.476 = 317.241$$

Substituting these values in the Taylor's expansion (A.2),

$$y(x) = 0.5 + 5.625x + 6.328x^2 + 52.87x^3 + \cdots$$

Number of terms used in the above equation depend on the accuracy of y required. There are some shortcomings in this method. It is not very accurate as sometimes large number of terms have to be considered and for higher order terms, derivatives become more and more cumbersome.

A.3 Polygon Method

Euler's method discussed in section A.1 is the simplest of all the one-step methods. It does not require any differentiation and is easy to implement on computers. However it's major problem is large truncation errors. *Polygon method* is an improvement of Euler's method and is discussed in this section.

In Fig. *A*.1, we see that y_{i+1} calculated by Euler's method is at point P' whereas it should actually be at point Q, which lies on the curve. Thus there is a truncation error in y equal to $P'Q$. This is because the tangent at initial point $T(x_i, y_i)$, meets the line $x = x_{i+1}$ at P' (Fig. A.1). The equation

$$y_{i+1} = y_i + hf(x_i, y_i)$$

in case of Euler's method, where $h = \Delta x$, and $f(x, y)$ is any function. Thus Polygon method is modification of Euler's method as

$$y_{i+1} = y_i + hf\left(\frac{x_i + x_{i+1}}{2}, \frac{y_i + y_{i+1}}{2}\right), \quad h = \Delta x$$

$$= y_i + hf(x_i + h/2, y_i + \Delta y/2) \tag{A.4}$$

Δy is the estimated incremental value of y from y_i and can be obtained using Euler's formula as

$$\Delta y = hf(x_i, y_i)$$

Thus equation (A.4) becomes

$$y_{i+1} = y_i + hf(x_i + h/2, y_i + hf(x_i, y_i)/2)$$

$$= y_i + hf(x_i + h/2, y_i + hm_1/2), \tag{A.5}$$

where

$$m_1 = f(x_i, y_i)$$

$$= y_i + m_2 h$$

where

$$m_2 = f(x_i + h/2, y_i + hm_1/2)$$

this method is called *Modified Euler's method or improved polygon method.* Sometimes it is also called *mid point method.*

Let us integrate equation

$$\frac{dy}{dx} = 2y/x$$

with initial conditions at $x = 1,\ y = 2.0$

by Polygon method and compared with its integration with Euler's method.

Now we assume that $h = 0.25$ then

$$y(1) = 2.0$$

$$y(1.25) = 2.0 + 0.25f(1 + 0.125,\ 2 + 0.125f(1,\ 2))$$

$$= 2.0 + 0.25\,f(1.125,\ 2.5) = 3.11$$

$$y(1.5) = 3.11 + 0.25f(1.25 + 0.125,\ 3.11 + 0.125f(1.25, 3.11))$$

$$= 3.11 + 0.2f(1.375,\ 3.732) = 4.47$$

Estimated value of $y(1.5)$ by various methods for the equation

$$\frac{dy}{dx} = 2y/x$$

with initial conditions at

$$x = 1,\ y = 2.0$$

are given by

Euler's method : 4.20

Polygon method : 4.47

Exact solution : 4.50

This shows that Polygon method gives results closer to exact solution. Below we give a C++ program for Polygon method.

Program A.2: Polygon Method

```
/* Computer program to integrate an ordinary differential equation by Polygon
Method*/
   #include <iostream.h>
   main()
```

```
{
int i,n;
float x,y,xp,h,m1,m2;
float func(float, float);
cout<<"\n Solution by polygon Method\n\n";
/* Reading Initial data*/
cout<<"\n Input Initial values of x and y\n";
cin>>x>>y;
cout<<"input x for which y is required\n";
cin>>xp;
cout<<"Input step size h\n";
cin>>h;
/*Compute number of steps required*/
n=(int)((xp-x)/h+0.5);
/*Compute y recursively at each step*/
for(i=1;i<=n;i++)
{
m1=func(x,y);
m2=func(x+0.5*h,y+0.5*h*m1);
x=x+h;
y=y+m2*h;
cout<<i<,x<<y<<"\n";
}
/*Write the final results*/
cout<<"Value of y at x="<<x<<"is"<<y<<"\n";
}//End of main()
float func(float x, float y)
{
float f;
f =2.0 *y/x;
return(f)
}
```

A.4 Runge-Kutta Method

In Polygon method, we have generated a second slope m_2 from slope m_1 of Euler's method as

$$m_1 = f(x_i, y_i)$$

$$m_2 = f(x_i + h/2, y_i + hm_1/2)$$

Now we discuss a method which is much more accurate as compared to above methods. This method is known as **Runge-Kutta method**.

This method further generates two more slopes i.e., m_3 and m_4 that is

$$m_1 = f(x_i, y_i)$$

$$m_2 = f(x_i + \frac{h}{2}, y_i + \frac{m_1 h}{2})$$

$$m_3 = f(x_i + \frac{h}{2}, y_i + \frac{m_2 h}{2})$$

$$m_4 = f(x_i + h, y_i + m_3 h) \tag{A.6}$$

$$y_{i+1} = y_i + h\left(\frac{m_1 + 2m_2 + 2m_3 + m_4}{6}\right)$$

These equations are called **Runge-Kutta method of fourth order**. For detailed derivation of the method readers may refer to [4].

A.5 Difference Equation for Partial Differential Equations

Following the logic in above section, now we extend it for partial differential equations of first and second degree. Now we consider a partial differential equation and convert it into a difference equation. The grid in Fig. A.1 can be considered as *t-x* plane in place of *x-y* plane. Let the equation be [4]

$$\frac{\partial u}{\partial t} + a(u,x,t)\frac{\partial u}{\partial x} = 0, \qquad 0 \le x \le 1,\ t > 0,\ a > 0 \tag{A.7}$$

where initial values are

$$u(x,0) = F(x), \quad 0 \le x \le 1$$
$$u(0,t) = g(t), \quad 0 \le t \le T$$

where $f(x)$ represents the initial conditions and $g(t)$ the boundary conditions at $x = 0$. No boundary conditions are required at $x = 1$.

A difference method might be set-up by selecting a grid with spacing Δx and Δt in the x and t direction respectively. Since in the present case there are two independent variables x and t, we use $u(x,t)$ as $u_j^n = u(j\Delta x, n\Delta t)$, where superscript is for n-th point along t-axis and subscript is for j-th point along x-axis in the grid. There are many ways of converting the equation (A.7) into difference equation. A popular choice for this simple transport equation is the following difference equation.

$$\frac{u_j^{n+1} - u_j^n}{\Delta t} + a(u_j^n, x_j, t^n)\frac{u_j^n - u_{j-1}^n}{\Delta x} = 0 \tag{A.8}$$

which can be solved at all positive integer values of j and n once values have been given for $j = 0$ and $n = 0$. As a special case if a is a constant, $f(x) = \sin(x)$ and $g(t) = -\sin(at)$, then the solution to the equation (A.7) is $u(x,t) = \sin(x - at)$. The difference equation (A.7) can be solved by letting $u_j^0 = f(j\Delta x)$ and letting $u_0^n = g(n\Delta t)$.

For detailed study of numerical techniques using difference equations, readers are referred to reference [4].

All numerical techniques for solving differential equations of the type $\frac{d^n y}{dx_n} = f(x, y)$ involve a series of estimation of $y(x)$ starting from the given conditions. There are two basic approaches that could be used to estimate the value of $y(x)$. They are known as *one-step methods* and *multiple step methods*.

In one step methods, we use the information from only one preceding point i.e., to estimate the value y_i, we need the condition at the previous point y_{i-1} only. Multi step methods use information at two or more previous steps to estimate a value. In this section, we will discuss some of the integration techniques for ordinary differential equations.

6

SIMULATION MODEL FOR AIRCRAFT VULNERABILITY

One of the applications of dynamic simulation is in the field of aircraft vulnerability studies. This is an important field of aircraft design industry. The study of vulnerability of a combat aircraft against ground based air defence system is of utmost importance for the design and development of aircraft, so that it is capable of surviving against it. Also such a software is used for the evaluation of various anti aircraft weapon system which are under development. This study is also needed in planning the Air Defence System (ADS) against an enemy attack. So far damage to stationary targets and mathematics behind it had been discussed in the previous chapters. In chapter three, it was assumed that projection of aircraft and its vital parts on a given plane have been provided as inputs to the model. Also effect of motion of aircraft is also ignored. In this chapter study of single shot hit probability to a moving target will be undertaken. For this purpose, aircraft will be taken as a moving target. Preliminary study of vulnerability, effect of redundancy and overlapping has already been given in chapter three. These concepts will be used in the present chapter. Number of models dealing with vulnerability of aerial targets, have been reported in the literature [2, 38, 51, 66–69]. A simplistic form of model where areas of vulnerable parts projected on a given plane are known, has been given by Ball [5] (see chapter three). There it was assumed in that the projection of aircraft as well as its vital parts on a typical surface is available as an input data. In the present chapter, we will give complete algorithm for the evaluation of vulnerability of a typical aircraft. Only inputs used in this model are the structural data of the aircraft and characteristics of weapon system.

A dynamic model of aircraft vulnerability due to ground air defence system, where aircraft is assumed to be moving along an arbitrary profile, is considered. A proximity fuse shell, fired from ground towards aircraft, has been assumed to explode in the near vicinity of the target aircraft. This shell has fragments as well as blast effects. If it explodes near the target, main damage may be due to blast only. Damage to aircraft body due to explosive charge as well as fragments, when warhead/ammunition explodes in its near vicinity has been considered in this chapter. Kill criterion has been taken as the minimum number of fragments required to penetrate and kill a particular part based on the total energy requirement. In the case of damage due to blast waves, it is assumed that the probability of kill is one, since energy released by shell is sufficient for catastrophic explosion. This assumption is based upon the total impulse transmitted to the aircraft structure. A typical aircraft and a typical air defence gun with DA/VT fused ammunition has been considered for the validation of the model. Structural data of aircraft as well as its vital parts is taken in the form of triangular elements [66]. These triangular elements are obtained by dividing surface of whole aircraft and it's vital parts into small three dimensional triangles. For this purpose one has to go to the drawings of the aircraft and obtain the (x, y, z) co-ordinates of apexes of all the triangles. Kill criterion due to fragment hits has been modified and is based on fragment energy concept. For DA fused impact, it has been assumed that shell first hits the outer surface of aircraft and then explodes. Kill in this case is mainly due to impact and explosive energy of the shell. A three dimensional model for single shot hit probabilities has been presented in this chapter for the case of proximity fused ammunition. The effect of redundant vulnerable parts on the overall kill probability of aircraft is also studied in this model.

6.1 MATHEMATICAL MODEL

In order to construct computer model of a typical aircraft whose survivability study is to be conducted, we divide the aircraft into large number of three-dimensional triangles, with co-ordinates of its apexes given by (x_i^k, y_i^k, z_i^k), where $i = 1, 2, 3$ and k is the number of the triangle. These triangles are arranged in required serial order and a three-dimensional aircraft is created in computer by perspective projection technique. This is done in order to counter check, that the data generated from drawings has no errors. Any error in data generation can easily be detected from the graphic output. A typical computer output of a typical aircraft has been given in Fig. 6.7. It has been assumed in the model that the aircraft is approaching towards a friendly vulnerable target (which is also the location of the air defence gun) in a level flight. The direction cosines of the aircraft's wind and body axes with respect to the fixed frame of reference with origin at the gun/missile position are given by aircraft flight profile equations. Once the aircraft enters the friendly territory, it is first detected by the surveillance radar and then is tracked by tracking radar, which completely gives its profile. Criterion for the kill of the aircraft is taken as, "aircraft is killed subject to some probability, if one of the vital part is killed. In Table 6.1, various levels of energies required to kill various vital parts is given whereas in Table 6.2, probability of kill of aircraft, subject to kill of a typical vital part has been shown.

Here it is important to know that the study of vulnerability of a combat aircraft against ground based air defence system is needed (*a*) to study the capability of an aircraft to survive against enemy attack during operations (aircraft combat survivability) and (*b*) to study the effectiveness of a newly developed weapon systems (missile or ground based guns).

For the evaluation of vulnerability of an aircraft, let us assume that location of the air defence gun (denoted by *G*) is the origin of a ground based co-ordinate system *G-XYZ* (here after to be called as frame-I). It is assumed that the aircraft is approaching towards the origin in a level flight. Since the structural data of aircraft as well as its vital parts is given in the form of triangular elements, in order to measure the co-ordinates of the triangles, a frame of reference fixed in the aircraft has been considered. Co-ordinates of all points on the aircraft has been measured in terms of co-ordinate system O-UVW (here after to be called frame-II) fixed in the aircraft, with origin *O* being the geometric centre of the aircraft. Thus, when observed from the point *G*, each and every point of aircraft is moving with speed of aircraft, where as it is stationary with respect to origin *O*. Kill criterion due to fragment hits is based on fragment energy concept. Fragment energy concept is explained in section 6.5.1. Damage to aircraft body due to explosive charge as well as fragments, when warhead/ammunition explodes in its near vicinity has been considered. Aircraft in fact is a complex system, in which each and every part has an important role to play. It has large number of vital components. But for the sake of simplicity only five vital parts are considered. These components are avionics, pilot 1, pilot 2, two fuel tanks and two engines. Fuel tanks and engines are taken as redundant vital parts. Fuel tanks are of two types; one is fuselage fuel tank and other is in two wings. Two fuel tanks located in two wings are treated as one part. Two pilot are assumed to be non-redundant parts.

Probability of kill of aircraft/vital part depends on probabilities of its detection, hit and fuse functioning. Probability of kill of j-th vital part of an aircraft by a round is defined as [66].

$$\overline{P}_k^j = P_d P^j{}_h P_f P^j{}_{k/hf} \qquad ...(6.1)$$

where P_d, P_h^j, P_f, $P^j{}_{k/hf}$ respectively are probabilities of detection of the aircraft, hit on the j-th vital part of the aircraft, fuse functioning and kill subject to being hit and fuse functioning. $P^j{}_h$ is hit on the j-th vital part, in case of DA fused ammunition and landing in the vicinity zone, in case of VT-fused ammunition. DA fuse ammunition is direct attack ammunition which hits the targets and explodes. Variable time fuse ammunition (VT fuse) explodes in close vicinity of the target. Vicinity zone is a region around the aircraft in which once shell lands, will explode with some probability. Probability of functioning is the design parameter of the fuse and varies from fuse to fuse.

Thus probability of kill of the aircraft, subject to kill of its vital parts, is given as

$$P_{k/hf} = 1 - \left(1 - P_{k,1}^1\right)\left(1 - P_{k,1}^2\right)\left(1 - P_{k,1}^3\right)\ \left(1 - P_{k,1}^4 P_{k,2}^4\right)\left(1 - P_{k,1}^5\ P_{k,2}^5\right) \qquad ...(6.2)$$

where $P_{k,i}^j$ is the kill probability of the i-th redundant part of j-th vital part. In this relation $P_{k,i}^j$ is same as $\overline{P}_k^j$ in relation (6.1), except that subscript i has been added and bar over P has been removed. This relation has been explained in chapter three in details.

6.2 Single Shot Hit Probability on N-plane

So far we have discussed the single shot hit probability on the targets which are two-dimensional i.e., areas. Since the aircraft is a three-dimensional body, we have to convert it into two-dimensional area. This is possible by finding its projection on a typical plane. Most appropriate plane is the plane normal to the line of sight. Thus we project the aircraft on the plane normal to the shot line. If the aircraft is stationary, single shot hit probability by a shell fired from the ground air defence gun can be evaluated in a manner similar to that given in section 2.10. In the present case aircraft has been assumed to be moving with respect to frame-I. In order to make it stationary with respect to this frame, three-dimensional data of aircraft, measured in frame-II will be transformed in terms of fixed frame co-ordinates with the help of linear transformations. Since from the gun position aircraft looks like a two-dimensional figure, which is nothing but the projection of the aircraft on a plane, normal to the line of sight, transformation of this data with respect to two-dimensional plane (N-plane, to be defined later) will be obtained. In frame-II of reference, point O is geometrical centre of the aircraft and OU, OV, OW are along the rolling, pitching and yawing axis (Fig. 6.1).

If A and E are respectively the angles in azimuth and elevation of the aircraft, measured from the ground based radar, then the direction cosines of line GO (line joining gun position and centre of the aircraft) are given by

$$l_0 = \cos A \cos E, m_0 = \sin A \cos E, n_0 = \sin E,$$

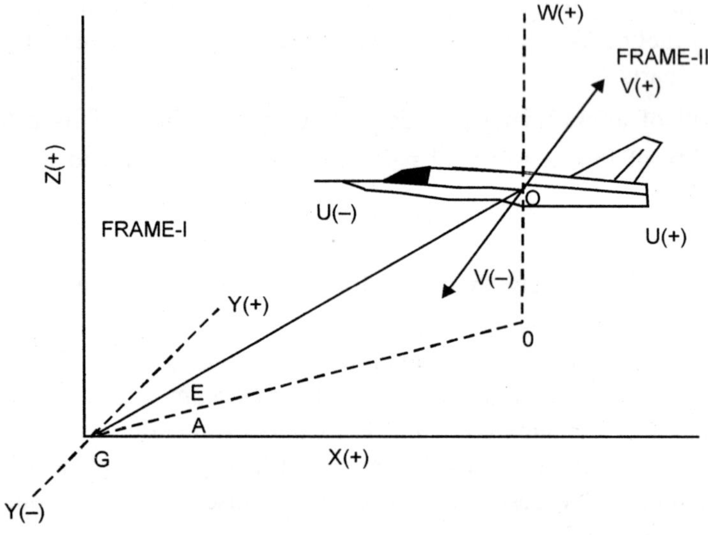

Fig. 6.1: *Different co-ordinate systems.*

Let co-ordinates of point P_i, which are apexes of the triangular elements be (u_i, v_i, w_i), in terms of O-UVW co-ordinate system. In order to achieve our goal (to find projection of aircraft on N-plane) we first find projection of a typical point P, of the aircraft on the N-plane. It is to be noted that point P is a function of time t. Thus co-ordinates of a typical point P at the aircraft in terms of fixed co-ordinate system (x, y, z) are obtained by linear transformations,

$$\begin{aligned} x_p &= x_0 + l_1 \cdot u_p + l_2 \cdot v_p + l_3 \cdot w_p \\ y_p &= y_0 + m_1 \cdot u_p + m_2 \cdot v_p + m_3 \cdot w_p \\ z_p &= z_0 + n_1 \cdot u_p + n_2 \cdot v_p + n_3 \cdot w_p \end{aligned} \quad ...(6.3)$$

where subscript p denotes co-ordinates of point P and (x_0, y_0, z_0) are the co-ordinates of the centre O of the aircraft. The direction cosines of line GP (point P is different from the point O, which is the centre of the aircraft), are

$$\begin{aligned} l_p &= x_p/GP \\ m_p &= y_p/GP \\ n_p &= z_p/GP \end{aligned} \quad ...(6.4)$$

and the angle between θ the lines GP and GO is

$$\theta = \cos^{-1}\left(l_0 l_p + m_0 m_p + n_0 n_p\right) \quad ...(6.4a)$$

where $GP = \sqrt{x_p^2 + y_p^2 + z_p^2}$ and (l_i, m_i, n_i), i = 1, 2, 3 are respectively direction cosines of *OU*, *OV*, *OW*, with respect to fixed frame *OXYZ*.

It is well known that when viewed from the ground, aircraft looks like a two-dimensional figure, which is nothing but its projection on a plane which passes through point *O* and is normal to the line of sight (line *GO*). This plane is called normal plane to the line of sight and is denoted by *N*-plane. Now consider a plane (*N*-plane) at right angles to *GO* passing through the point *O* (Fig. 6.2). Our aim is to find projection of aircraft on this plane. In order to achieve this, first we will find the projection of a typical point *P* of the aircraft on *N*-plane. Let this projection be a point *Q*. Locus of point *Q* will be nothing but the projection of the aircraft on *N*-plane. To get the co-ordinates of point *Q*, we convert three-dimensional co-ordinates of the point *P* in terms of two-dimensional co-ordinates of point *Q* in *N*-plane. Let these co-ordinates be (s_q, t_q) in *O-ST* axis in *N*-plane defined below. Consider a two-dimensional co-ordinate frame *O-ST* in the *N*-plane such that *OT* is in the vertical plane containing line *GO* and *OS* in the horizontal plane through *O*. Then the direction cosines of the *OS*-axis with respect to the *GXYZ* frame are (appendix 6.1)

$$\left(\frac{m_0}{1-n_0^2}, \frac{-l_0}{\sqrt{1-n_0^2}}, 0\right) \approx \left(l'_s, m'_s, n'_s\right) \qquad \ldots(6.5)$$

and of *OT* axis are

$$\left(\frac{-l_0 n_0}{\sqrt{(1-n_0^2)}}, \frac{-m_0 n_0}{\sqrt{(1-n_0^2)}}, \sqrt{(1-n_0^2)}\right)$$

The point *Q* at which the line *GP* (produced, if necessary) meets the *N*-plane satisfies the equations,

$$GQ = GO/\cos\theta; \; OQ = GO \tan\theta$$

Since line *GQ* is nothing but the extension of line *GP*, therefore co-ordinates of the point *Q* in the *GXYZ* frame turns out to be

$$x_q = GQ \cdot l_p, \; y_q = GQ \cdot m_p, \; z_q = GQ \cdot n_p$$

where (l_p, m_p, n_p) have been derived in equation (6.4).

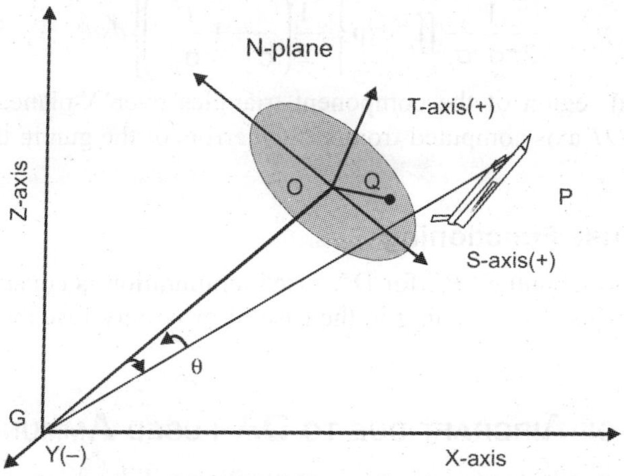

Fig. 6.2: *Projection scheme on N-plane.*

Thus, the direction cosines of the line OQ (which is line on the N-plane) with respect to the G-XYZ frame are

$$l_q = \left(x_q - x_0\right)/OQ$$

$$m_q = \left(y_q - y_0\right)/OQ$$

$$n_q = \left(z_q - z_0\right)/OQ$$

where $OQ = \sqrt{(x_q - x_0)^2 + (y_q - y_0)^2 + (z_q - z_0)^2}$. Finally, the co-ordinates (s_q, t_q) of the point Q in the N-plane are given by

$$s_q = OQ \cdot \cos\phi$$

$$t_q = OQ \cdot \sin\phi$$

where ϕ is the angle, line OQ makes with OS axis in N-plane and is given by,

$$\phi = \cos\left(l_q \cdot l'_s + m_q \cdot m'_s + n_q \cdot n'_s\right)n$$

where (l'_s, m'_s, n'_s) are the direction cosines of the OS-axis with respect to G-XYZ frame. Derivation of (l'_s, m'_s, n'_s) is given in appendix 6.1.

6.2.1 Single Shot Hit Probability

Let F_p be the shape of a typical part of the aircraft body bounded by the line segments with vertices P_i (i = 1, 2,..., n), then corresponding points Q_i(i = 1, 2,..., n) of the projection of the part on N-plane can be determined as explained above and a corresponding projection F_q can be obtained. The projection F_q is such that a hit on this area will imply a hit on the part F_p of the aircraft body. Similar analogy can be extended for other parts of the aircraft even those which are bounded by the curved surfaces.

Thus single shot hit probability on the part F_p of the aircraft is given by,

$$P_h = \frac{1}{2\pi\sigma_s\sigma_t}\iint_{F_q} \exp\left\{-\frac{1}{2}\left(\frac{s^2}{\sigma_s^2} + \frac{t^2}{\sigma_t^2}\right)\right\} ds dt \qquad ...(6.6)$$

where F_q is the projected region of the component triangles over N-plane, σ_s, σ_t, are the standard deviations along OS and OT axis computed from system errors of the gun in the azimuth and elevation plans respectively.

6.2.2 Probability of Fuse Functioning

The probability of fused functioning "P_f" for DA fused ammunition is constant and is taken as a part of the data. Probability of fused functioning in the case of proximity fuse (VT fuse) will be discussed in the coming sections.

6.3 Vulnerability of Aircraft due to DA-Fused Ammunition

In this section, kill of an aircraft due to DA fused ammunition will be discussed. By DA fuse we mean, a fuse which only functions when shell hits the target. In order to determine the damage caused by the AD guns (fitted with DA fuse) to a given aircraft, following criteria have been adopted in the model.

(*a*) Considering the actual terminal velocity, mass and calibre of shell, penetration in the vital as well as non-vital parts has been calculated.

(*b*) If the energy released by a shell is greater than the energy required (Table 6.1) to kill a vital part, then probability of kill of that part is taken as 1.0 and probability of kill of aircraft is calculated as per the Table 6.2.

DA fuse has an in-built delay mechanism. This is due to the fact that , shell first has to penetrate the target and then explode. Kinctic energy of the projectile is responsible for the penetration of the projectile in the vulnerable parts. When a projectile penetrates a surface, its velocity and mass both reduce and thus total kinetic energy reduces. This is due to the resistance offered by the target. In the case of aircraft (and same is true in case of other targets too) damage to a vital part is caused due to the remaining energy which is left with the projectile after crossing the outer skin. Remaining velocity of projectile after penetration in the vulnerable part is given by [75]

$$V_r = \left(V - x\rho D_0 R^2 V \sin\alpha / m\right)\cos\theta \qquad ...(6.7)$$

where

V = normal striking velocity,

V_r = remaining velocity,

ρ = density of target,

α = nose cone angle,

θ = striking angle of projectile,

D_0 = thickness of the target,

R = radius of projectile,

m = mass of projectile.

In the present model it has been assumed that if shell hits at any of the vital or non-vital (rest of aircraft body) part, its kill probability depends on the factor, whether shell penetrates the aircraft body or not. The kill probabilities given in Table 6.2 are for the *K*-kill (Chapter three) of aircraft, subject to kill of vital part, when a small calibre impact fused high explosive projectile (in the present case 23 mm shell) hits it. If shell is of higher calibre which release energy Q then these kill probabilities are multiplied by factor

$$E_f = 2.0(1 - 0.5\exp(-(Q - Q_1)/Q_1)) \qquad ...(6.8)$$

where Q_1 is the energy released by a typical 23 mm shell with 20 gm HE. This has been done in order to take the effect of higher explosive energies of different shells on the kill of aircraft. It is well known that damage to a target is an exponential function of explosive energy. Thus probability of kill of *j*-th component $\left(P^j_{k/h}\right)$ is given as,

$$P^j_{k/i} = P^j_{k/h} F^j_{k/i} \cdot E_f \qquad ...(6.8a)$$

where $P^j_{k/h}$ is the probability of kill given by equation (6.17), and factor $F^j_{k/i}$ is

Table 6.1: Equivalent thickness of various vital and non-vital parts

Components	Energy required to kill vital part (J)	Equivalent thickness of dural (mm)
1. Avionics	339	27 mm
2. Pilot sec.	678	5 mm
3. Engine	1356	20 mm
4. Fuel tanks	339	20 mm
5. Remaining parts	400	5 mm

Table 6.2: Probability of K-type kill for various vital components

Components	Probability of K-kill $F^j_{k/i}$	
	Internal burst %	External burst %
1. Avionics	50	10
2. Pilot	50	10
3. Engines	35	25
4. Fuel tanks	75	30
5. Remaining parts	25	05

given in the Table 6.2. It is to be noted that factor E_f is multiplied with the total kill probability only in the case of DA fused ammunition. In equation (6.8*a*), *j* is not the dummy index but indicates *j*-th vital part.

6.4 Probability of Landing in Case of Proximity Fuzed Shell

Proximity fuse (PF) is different from DA fuse in a sense that it detects the target and functions with some probability. Probability of fuse functioning depends upon the distance from the target and is given by,

$$P_{df}(r) = \frac{1}{C} P_f(r) \qquad ...(6.9)$$

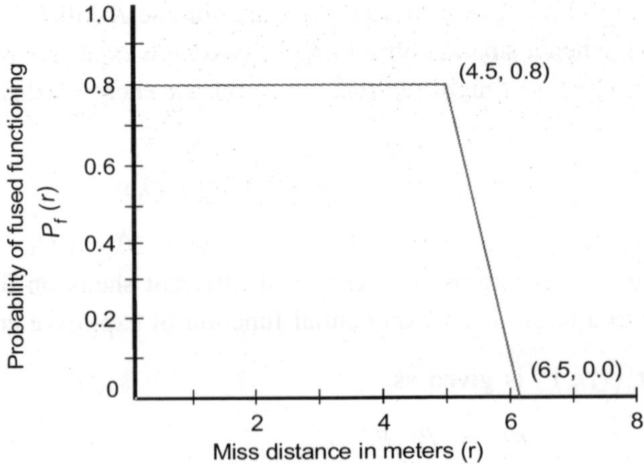

Fig. 6.3: *Probability of fused functioning.*

where $C = \int_0^7 P_f(r)\,dr$

and

$$\begin{aligned} P_f(r) &= 0.8, \text{ for } r \leq 4.5 \\ &= 0.4r + 2.6, \text{ for } 4.5 \leq r \leq 6.0 \\ &= -0.2r + 1.4, \text{ for } 6.0 \leq r \leq 6.5 \\ &= 0.0, \text{ for } r > 6.5 \end{aligned}$$

where r being the normal distance from the surface of the target.

Probability of fuse functioning in the above equation is for a typical fuse (Fig. 6.3) and may be different for different types of fuses. In case of proximity fused ammunition shell can land anywhere in the vicinity region of the aircraft.

Vicinity region of a target is a region around it, so that if a shell lands within it, its fuse functions. Probability of fuse functioning depends on the normal distance from the surface of the target. Let m_s be the maximum distance at which there is a probability that fuse will function. This distance is called miss distance i.e., the distance beyond which if shell lands, fuse will not function. In fact, vicinity region has an irregular shape depending on the reflected signals coming from different parts of the aircraft. To model such a region mathematically is a tedious task. Thus we have assumed that this region is of cylindrical shape with its axis, same as the axis of the aircraft. Consider a cylinder with axis along the axis O-U of the aircraft, whose radius is $R_v = R_a + m_s$ and length $2(l_0 + m_s)$ mid-point of cylinder, being the geometric centre of aircraft, R_a and m_s being the radius of the aircraft and miss-distance.

Probability of landing and fuse functioning of projectile in terms of fixed co-ordinate system, around aircraft is

$$P_{Lf} = \frac{1}{\left(\sqrt{2\pi\sigma}\right)^3} P_f(r - R_a) \cdot \exp\{-(\bar{z})^2 + (\bar{y})^2 + (\bar{z})^2 / 2\sigma^2\} d\bar{x}\, d\bar{y}\, d\bar{z} \qquad \text{...(6.10)}$$

where

$$\bar{x} = x - x_0,$$

$$\bar{y} = y - y_0,$$

$$\bar{z} = z - z_0$$

and $P_f(r - R_a)$ is the probability of fuse functioning. R_a is value of r at the surface of aircraft, r being the radial distance from the u-axis. Relation (6.10) is a simple extension of two dimensional Gaussian distribution to three dimensional case. Converting $(\bar{x}, \bar{y}, \bar{z})$ co-ordinate, to moving co-ordinates (u, v, w) with the help of linear transformations (inverse of equations 6.3) one gets,

$$P_{Lf} = \frac{1}{(\sqrt{2\pi}.\sigma)^3} P_f(r - R_a) \cdot \exp\left\{-\frac{(u^2 + v^2 + w^2)}{2\sigma^2}\right\} J_1 \left(\frac{\bar{x}, \bar{y}, \bar{z}}{u, v, w}\right) du \cdot dv \cdot dw$$

where

$$J_1\left(\frac{\bar{x},\bar{y},\bar{z}}{u,v,w}\right) = \begin{vmatrix} l_1 & l_2 & l_3 \\ m_1 & m_2 & m_3 \\ n_1 & n_2 & n_3 \end{vmatrix}$$

is the Jacobian, used for transformation of co-ordinates from one set of co-ordinate system to other set of co-ordinate system. Transforming above equation to cylindrical co-ordinates (r, θ, ζ) one gets,

$$P_{Lf} = \frac{1}{\left(\sqrt{2\pi\sigma^2}\right)^3} P_f\,(r - R_a)\cdot\exp\left\{-1/2\frac{r^2+\zeta^2}{\sigma^2}\right\} J\cdot r\cdot d\zeta\cdot dr\cdot d\theta \qquad \text{...(6.11)}$$

The probability of kill of j-th vital part due to shell landing at a typical point $P\,(r,\theta,\zeta)$ of vicinity shell is,

$$dP_k^j = P_{Lf} P_{k/h}^j \qquad \text{...(6.12)}$$

where $P_{k/h}^j$ is the kill probability of j-th part subject to a hit on it (6.8a). Since the shell can land any where in the vicinity region, the cumulative kill probability $P_{k/i}^j$ of j-th part, due to i-th (say) shell landing anywhere in the vicinity of the aircraft is obtained by integrating (6.12) over the whole vicinity shell i.e.,

$$P_{k/i}^j = \left(\int_{R_a}^{R_L} + \int_{R_L}^{R_v}\right)\int_0^{2\pi}\int_{-(l_0+m_s)}^{(l_0+m_s)} P_{Lf}\cdot P_{k/h}^j\cdot r\,dr\,d\theta\,d\zeta \qquad \text{...(6.13)}$$

Expression for $P_{k/h}^j$ will be derived in coming sections. Evaluation of $P_{k/h}^j$ depends on the kill criteria. In integral (6.13), limit R_L is a typical distance from the aircraft such that if shell explodes between R_v and R_L, damage is due to blast as well as fragments, otherwise it is only due to fragments. In the next section, evaluation of parameter R_L has been discussed.

6.4.1 Determination of R_L

The estimation of R_L can be done on the basis of critical 'impulse failure criterion'(page 134 of [14]). This criterion essentially states that structural failure under transient loading can be correlated to critical impulse applied for a critical time duration where the latter is assumed to be one quarter of the natural period of free vibration of the structure. This critical impulse can be expressed as:

$$I_0 = (\rho/E)^{1/2}\cdot t\cdot\sigma_y \qquad \text{...(6.14)}$$

where

E = Young's modulus of the target material,
ρ = density of material,
t = thickness of the skin,
σ_y = dynamic yield strength.

Aircraft structure is a combination of skin panels supported by longitudinal and transverse membranes. In applying this method to skin panels supported by transverse and longitudinal membrane, one first calculates the critical impulse of the blast wave interacting with the panel and the natural period of the panel. Incident pressure pulse having a duration of one quarter of the natural period or more having an impulse at least equal to I_c will rupture the panel at that attachment.

Incident pressure of a blast wave is a function of distance from the target as well as total energy released by the shell. If the scaled distance of point of explosion from the target is z then [35]

$$P^0/P_a = \frac{808\left[1+(z/4.5)^2\right]}{\sqrt{1+(z/0.048)^2}\sqrt{1+(z/0.32)^2}\sqrt{1+(z/1.35)^2}}$$

where

P^0 = incident blast over pressure,

P_a = atmospheric pressure,

$z = R/W^{1/3}$,

R being the distance from the point of explosion and W the weight of explosive.

A typical blast pulse has been shown in Fig. 6.4. Since the detailed study of shock wave is not the subject of this book, readers may go through some book on shock waves [35, 57], to know more about this subject.

Time duration (t_d) of positive phase of shock is given by the following relation,

$$\frac{t_d}{W^{1/3}} = \frac{980\left[1+(z/0.54)^{10}\right]}{\left[1+(z/0.02)^3\right]\left[1+(z/0.74)^6\right]\sqrt{1+(z/6.9)^2}}$$

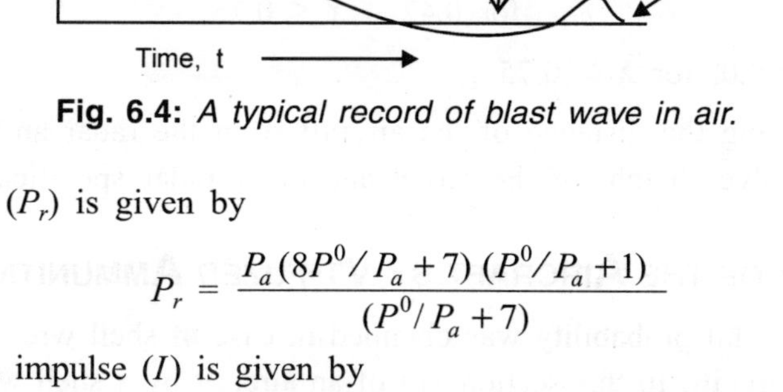

Fig. 6.4: *A typical record of blast wave in air.*

and reflected pressure (P_r) is given by

$$P_r = \frac{P_a(8P^0/P_a+7)(P^0/P_a+1)}{(P^0/P_a+7)}$$

Therefore the total impulse (I) is given by

$$I = \int_0^{t_d} P_r\,dt$$

where t_d is the duration of the shock pulse. If we assume the reflected pressure pulse (which is quite fair approximation) to be a triangular pressure pulse, then

$$I = \frac{P_r \cdot t_d}{2}$$

Taking the dimensions of the panel as a and b, the natural frequency (ω) of fundamental mode is [66]

$$\omega = \pi^2 \left(1/a^2 + 1/b^2\right) \sqrt{\frac{Et^3}{12(1-\nu^2)\rho}}$$

where

E = Young's modulus,

ρ = density,

t = thickness,

ν = Poisson's ratio.

Then the natural time period T, of panel is $T = 2\pi/\omega$.

Keeping in view the above relations, we can simulate the value of z for which $I > I_C$. This simulated value of z will be equal to RL.

6.4.2 Probability of Detection by Radar

In chapter three, probability of detection by Radar was discussed using Radar equations. Probability of detection of aircraft is an important parameter for the assessment of its survivability/vulnerability, and for a typical air defence radar. Sometimes probability of detection, in the case of certain radar is given as a function of distance of the target. In this case knowledge of various parameters of radar is not needed. Probability of detection due at typical radar is given as,

$$P_d(X) = 1.0, \text{ for } 0 \le X < 0.15$$

$$P_d(X) = 31.470X^3 - 33.7136X^2 + 8.57498X + 0.33782, \quad \text{for } 0.15 \le X \le 0.42 \qquad \ldots(6.15)$$

$$P_d(X) = -8.269X^3 + 18.579X^2 - 13.9667X + 3.51733, \quad \text{for } 0.42 < X \le 0.75$$

$$P_d(X) = 0.0, \text{ for } X > 0.75$$

where $X = (R/R_0)$, R being the distance of the aircraft from the radar and for the typical radar, $R_0 = 65.58$ km. R_0 involves height of the target and other radar specifications [32].

6.5 Vulnerability of the Aircraft by VT-Fuzed Ammunition

In section 6.4 single shot hit probability was obtained in case of shell with proximity fuse landing in the near vicinity of aircraft. In this section kill of an aircraft by a shell with proximity fuse will be discussed. First we will assume that if k number of fragments with total energy E_r, penetrate a vital part, it will be killed. Later this concept will be amended suitably to obtain most appropriate kill.

It has been assumed that an aircraft is assumed to be killed with some probability, if at least one of its vital parts is killed. A vital component is treated as killed if the remaining energy of the fragment penetrating the vital part is more than the energy required to kill it (KILL CRITERION). It is quite possible that a fragment may not have sufficient energy required to kill a vital part. Singh and Singh [66] have assumed that if at least k number of fragments whose total kinetic energy is equal to the required energy E_r (Table 6.1), should penetrate a part in order to achieve its kill. Thus probability of kill P_{km}^{i}, that at least k number of fragments, whose mass is greater than m, should penetrate is given by Poisson's law as,

$$P_{km}^{i} = 1 - \sum \sum_{N=0}^{k-1} \frac{(m_r)^N}{N!} e^{-m_r} \qquad \text{...(6.16)}$$

where m_r is the average number of fragments penetrating the component. In this assumption there is a weakness. If a fragment hits but does not penetrate, may be due to the fact that it does not have sufficient energy for doing so, then in such a situation it will not cause any damage to the component. This mean fragment should have sufficient energy to damage the component. If a fragment does not have sufficient energy required for penetration in aircraft's skin, it will be reflected from its surface. Even if there are ten such fragments will not be able to damage the aircraft. Keeping this in mind this criterion later was modified by another criterion, known as energy criterion and is being discussed below.

6.5.1 Energy Criterion for Kill

In equation (6.16) there is one drawback. Although total energy of k fragments may be equal to E_r but their individual energy is so low that it may not penetrate even the outer skin. Thus from this, one can derive an idea that energy of each fragment has to be more that a minimum energy (to be called uncritical energy) required to penetrate the outer skin. Thus the above concept of kill is modified as follows. Thus the probability of kill of vital part in case of non exploding projectiles *viz*., fragments can be defined as,

$$P_{km}^{i} = \begin{cases} 1, E_r > E_c \\ \dfrac{E_r - E_u}{E_c - E_u}, E_u < E_r < E_c \\ 0, E_r < E_u \end{cases} \qquad \text{...(6.17)}$$

where E_r is the kinetic energy of a fragment after penetrating the outer structure.

E_u is the uncritical energy so that if energy imparted to aircraft component is less than E_u, no damage is caused, E_c, is the critical energy required to kill the component, so that if energy imparted to aircraft is more than E_c complete damage is caused. This equation is used for each and every fragment and then cumulative kill due to all the fragments is evaluated. For this, knowledge of expected number of fragments hitting a part is needed, which will be explained in the following section.

6.6 Expected Number of Fragment Hits on a Component

Let a shell bursts at point P^k in the vicinity region of the aircraft. Fragments of the shell after explosion move in the conical angular zones with respect to the axis of the shell. Let there are n_z such uniform conical zones, so that five degree is the apex angle of each zone. These zones are called uniform because average number of fragments per unit solid angle is constant with in a particular zone.

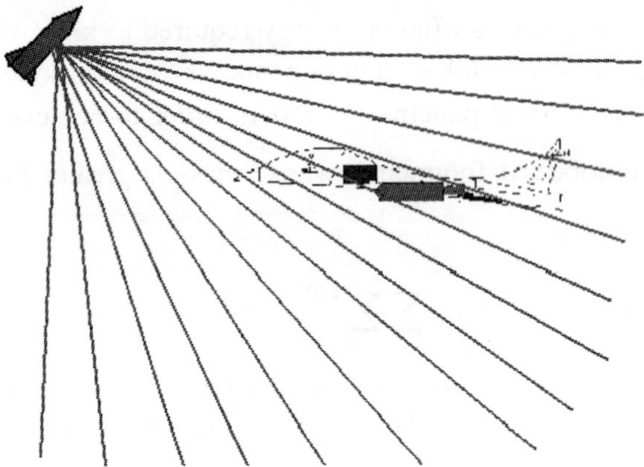

Fig. 6.5: *Pictorial view of a shell exploding in the near vicinity of aircraft. (Straight lines emerging from shell centre are fragment zones.)*

Since the size of the shell is very small as compared to that of aircraft, it can be assumed that fragments are being ejected as if they are coming from a point, which is centre of the shell (Fig. 6.5).

We define, $f^k_{i,i+1}$ as the region, which is the intersection of two solid cones, with vertex at point P^k and whose slant surfaces make angles α_i, α_{i+1} respectively with the axis of the shell, positive direction being towards the nose of the shell. Let $n^k_{i,i+1}$ be the total number of fragments of mass greater than m, in the angular zone $n^k_{i,i+1}$. Fragments per unit solid angle in the $z^k_{i,i+1}$-th angular zone can be given as

$$f^k_{i,i+1} = \frac{n^k_{i,i+1}}{2\pi(\cos\alpha_i - \cos\alpha_{i+1})} \quad ...(6.18)$$

It is to be noted that angles α_i, α_{i+1} are the angles of fragment cones in static mode i.e., when shell has no velocity. Distribution and number of fragments after explosion are obtained experimentally by exploding a static shell in air and measuring the distribution of fragments by collecting them in cardboard cabin made for the purpose. But it is known that in actual practice shell is moving with reference to fixed co-ordinate system and these angles have to be evaluated in that mode. Let these angles in dynamic mode be $\alpha'_j \cdot \alpha'_{j+1}$ (see relation 6.22).

Let $\omega^k_{i,i+1}$ be the solid angle subtended by the component in $z^k_{i,i+1}$ the angular zone. The total number of fragment hits on a component from this zone will be given by

$$N^k = \sum_{i=1}^{n_z} \omega^k_{i,i+1} \cdot f^k_{i,i+1} \quad ...(6.19)$$

These are the average number of fragments hitting a vital part, which is used in equation (6.16) and is denoted by a symbol m_r.

The solid angle subtended in the angular zone $z_{i,i+1}$ (here we have omitted the superscript k as these equations hold good for all k's) by a component at the centre of gravity of the shell is determined by the intersecting surface of the component and the angular zone $z_{i,i+1}$, and can be given by (Fig. 6.6)

$$\omega_{i.i+1} = \sum_{A_{i,i+1}} \delta w \qquad \text{...(6.20)}$$

where

$$\delta w = \frac{|\cos\theta|\delta A}{R_A^2} \qquad \text{...(6.20}a\text{)}$$

where

$A_{i,i+1}$ = intersecting surface of the component and the zone $z_{i,i+1}$ which will differ in stationery and dynamic cases.

A = a small elemental area on the vital component whose surface area is $A_{i,i+1}$.

R_A = distance between centre of gravity of the shell and mid point of δA.

θ = angle between R_A and normal to the surface at the mid point of δA.

Value of δw is evaluated in equation (6.20a). Following is the example to evaluate the solid angle subtended by a component of the aircraft in the different angular zones of the PF-shell, when the shell bursts at any arbitrary point C_s in the vicinity region of the aircraft. Using this method kill for any component of the aircraft can be obtained, having well defined surface.

Let the PF-fused shell bursts at a point C_s (also it is centre of gravity of the shell) in the vicinity region of the aircraft say at time $t = 0$. Let the co-ordinates of the centre of gravity of the shell at time $t = 0$ is (x_s, y_s, z_s) and velocity is v_s in the direction of (l_s, m_s, n_s) which is the direction of its axis with respect to I-frame, fixed in space.

Further let at the time of burst, (x_a, y_a, z_a) are the co-ordinates of the centre of the aircraft which is also the origin of the II-frame and let,(l_i, m_i, n_i), $i = 1, 2, 3$ are the direction cosines of the aircraft's axes (i.e., axes of the second frame) with respect to frame-I and this aircraft (or frame-II) is moving with velocity V_a in the direction (l_v, m_v, n_v) in the frame-I.

If co-ordinates of centre of shell at the time of burst ($t = 0$) with respect to frame-II of reference are (u_s, v_s, w_s), then transformation from one system of co-ordinates to other is given as

$$x_s = u_s \cdot l_1 + v_s \cdot l_2 + w_s \cdot l_3 + x_a$$

$$y_s = u_s \cdot m_1 + v_s \cdot m_2 + w_s \cdot m_3 + y_a$$

$$z_s = u_s \cdot n_1 + v_s \cdot n_2 + w_s \cdot n_3 + z_a$$

$$u_s = x_s \cdot l_1 + y_s \cdot m_1 + z_s \cdot n_1 + x_a$$
$$v_s = x_s \cdot l_2 + y_s \cdot m_2 + z_s \cdot n_2 + y_a$$
$$w_s = x_s \cdot l_3 + y_s \cdot m_3 + z_s \cdot n_3 + z_a \quad (6.21)$$

Let us assume that PF-fused shell bursts in stationary position with reference to frame-I α_i, α_{i+1} are the angles which the boundaries of the conical angular zone of fragments $z_{i,i+1}$ make with the positive direction of the shell axis and $V_{fi}, V_{f(i+1)}$ are the corresponding velocities of the fragments of these boundaries.

When shell bursts in dynamic mode, the directions and velocities of fragments, as observed in frame-I will be,

$$\alpha_i' = \tan^{-1}(V_2 / V_1)$$
$$V_{fi}' = \left(V_1^2 + V_2^2\right)^{1/2} \quad ...(6.22)$$

where

$$V_1 = V_s + V_{fi} \cos(\alpha_i)$$

$$V_2 = V_{fi} \cdot \sin(\alpha_i)$$

Fragments emerging from the point C_s in an angular zone $z_{i,\,i+1}$ will be confined in a cone making angles α'_i and α'_{i+1} respectively with the axis of the shell. Intersection of this cone with component is say an area P_1, P_2, P_3, P_4. Divide the surface enveloping P_1, P_2, P_3, P_4 into a finite number of rectangular areas $\delta A = \delta l \cdot \delta b$ (say) (Fig. 6.6) where δl and δb are dimensions of the rectangular element.

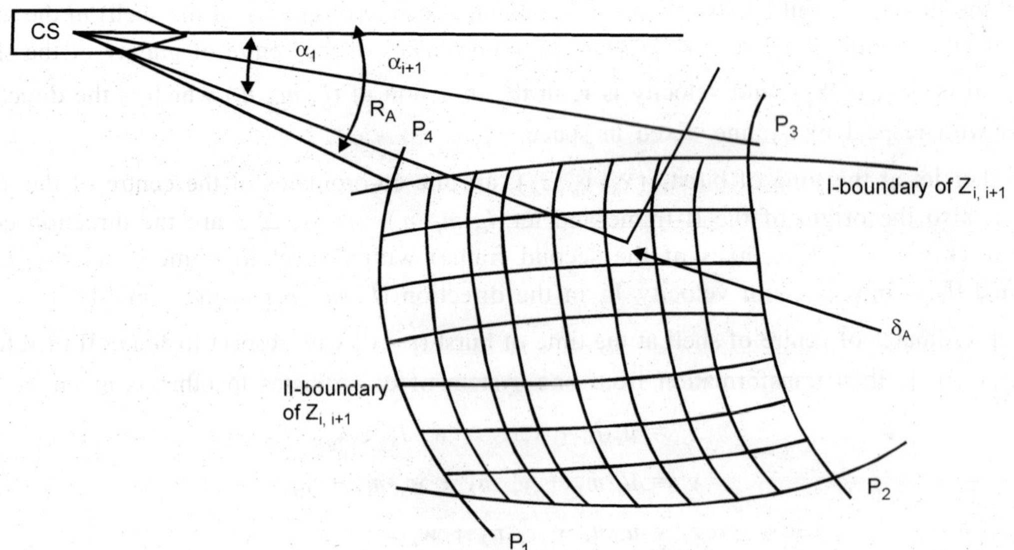

Fig. 6.6: *Interaction of a fragment with an element of aircraft.*

If a typical point P whose co-ordinate with respect to frame-II are $\left(u_p, v_p, w_p\right)$ is the middle point of area δA, then solid angle of area δA subtended at C_s and angular zone to which it belongs is determined by simulation as follows.

Co-ordinates of point P with reference to the fixed frame, at time t after burst are

$$x_{pt} = x_p + V_a \; l_v \cdot t$$
$$x_{pt} = y_p + V_a \; m_v \cdot t \qquad \text{...(6.23)}$$
$$z_{pt} = z_p + V_a \; n_v \cdot t$$

If θ is the angle between shell axis and the line $C_s - P_t$ where point $P_t(x_{pt}, y_{pt}, z_{pt})$ being the position of P at time t, then first step is to determine the angular zone α'_i, α'_{i+1} in which θ lies.

Fragment may come to the point P_t from angular zone $z_{i,\,i+1}$ with velocity $F'_{fi}\; F'_{f(i+1)}$ depending upon is θ close to.

Distance travelled by the fragment along the line $C_s - P_t$ in time t

$$D_f = V_f' \cdot t \qquad \text{...(6.24)}$$

where V_f' = selected $\left(V_{fi}' \, V_{f(i+1)}'\right)$. This means V_f' takes the value V_f', or $V_{f(i+1)}'$ subject to the condition that it is near to α_i' or α_{i+1}'. Actual distance between C_s *and* P_t is

$$D_s = \left[\Sigma\left(x_s - x_{pt}\right)^2\right]^{1/2} \qquad \text{...(6.25)}$$

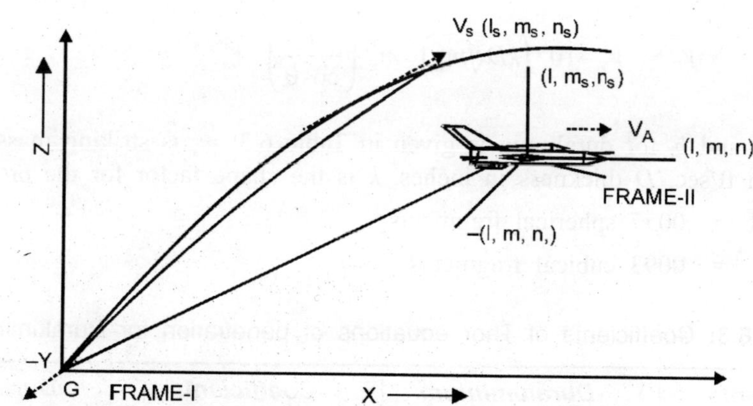

Fig. 6.7: *Exploding of shell near aircraft.*

From equations (6.24) and (6.25) we simulate time t, such that $D_f = D_s$ for confirmed impact. The velocity of impact to the aircraft, V_{strike} (relative impact velocity) is,

$$V_{\text{strike}} = \left(V'^2_f + V_a^2 - V_f' V_a \cos\beta\right)^{1/2} \qquad \text{...(6.26)}$$

where β is the angle between the positive direction of aircraft's velocity vector and fragment's velocity vector. Penetration for strike velocity V_s can be obtained from experimental data [14]. Penetration depends on angle of impact. If θ is the angle of impact, then

$$V_\theta = V_0/\cos\theta \qquad ...(6.27)$$

where V_θ is the required velocity of impact at angle θ when velocity at zero angle of impact is given.

If velocity of impact V_{strike} is such that it is more than some critical velocity then the solid angle δω, subtended by the small rectangular element, in the angular zone $z_{i,i+1}$, given by

$$\delta\omega = \frac{dA\cdot|\cos\theta|}{D_s^2} \qquad ...(6.28)$$

is added to the relation (6.20).

6.7 Penetration Laws

In the above section it was assumed that penetration of the fragments has been taken from the experimental data. In this section we will use actual penetration equations to achieve this result. The fragments before hitting the component penetrates the aircraft structure, if component is hidden inside the structure. After it penetrates the outer skin, some of the energy will be lost. The remaining energy after penetration will be responsible for the damage to the component. Similarly, mass of the fragment as well as its velocity will also decrease. The remaining mass m_r and remaining velocity V_r of fragment is given by Thor equations as [2],

$$m_r = m_s - 10^{a_1}\left(kD\sqrt[3]{m_s^2}\right)^{a_2} m_s^{a_3}\left(\frac{1}{\cos\theta}\right)^{a_4} V_s^{a_5}$$

$$V_r = V_s - 10^{b_1}\left(kD\sqrt[3]{m_s^2}\right)^{b_2} m_s^{b_3}\left(\frac{1}{\cos\theta}\right)^{b4} V_s^{b_5} \qquad ...(6.29)$$

where values of a_i and b_i for duralumin is given in Table 6.3. m_s is striking mass in grains, V_s is striking velocity in ft/sec, D thickness in inches, k is the shape factor for the projectile.

Shape factor k = .0077 spherical fragments

= .0093 cubical fragments

Table 6.3: Coefficients of Thor equations of penetration for Duraluminium

Coefficients	*Duraluminium*	*Coefficients*	*Duraluminium*
a_1	–6.663	b_1	7.047
a_2	0.227	b_2	1.029
a_3	0.694	b_3	–1.072
a_4	–0.361	b_4	1.251
a_5	1.901	b_5	–0.139

Table 6.4: Critical and uncritical energies for various components

	Uncritical energy (in joules)	***Critical energy (in joules)***
(i) Avionics	81.4	339.0
(ii) Pilot	81.4	678.0
(iii) Fuel tanks	81.4	339.0
(iv) Engine	135.6	1356.0

6.8 CUMULATIVE KILL PROBABILITY

So far we have studied the kill probability of a single component due to fragments /shell. As the aircraft is considered to have been divided into y parts, let $P^j_{k/i}$ be the single shot kill probability of a j-th vital part due to i-th fire, each fire having n rounds. The cumulative kill probability $\overline{P}^j_k$ j-th vital part in N number of fire can be given as

$$\overline{P}^j_k = 1-\prod_{i=1}^{N}\left[1-P^j_{k/i}\right]^n \qquad ...(6.30)$$

Further the aircraft can be treated as killed if at least one of its vital parts is killed. Thus in this case the cumulative kill probability for the aircraft as a whole can be given as:

$$\overline{P}_k = 1-\prod_{i=1}^{y}\left[1-P^j_k\right] \qquad ...(6.31)$$

where factor $P^j_{k/l}$ is the probability of kill of aircraft subject to a vital component kill (Table 6.2 and eqn. (6.8a)). In case the j-th part of the aircraft has y_i redundant parts and $\overline{P}$ is given by

$$\overline{P}_k = 1-\prod_{j=1}^{y}\left[1-\prod_{i=1}^{y}\left[1-P^j_{k/r}\right]\right] \qquad ...(6.31a)$$

where $P^j_{k/r}$ denotes the kill probability of the r-th redundant part of the j-th vital part.

6.9 DATA USED

For the evaluation of this model by numerical integration of equations, two types of inputs are needed i.e.,

- Structural data of the aircraft in terms of triangular elements
- Vital parts of aircraft are:
  - (*i*) Avionics
  - (*ii*) Pilot 1
  - (*iii*) Pilot 2
  - (*iv*) Fuselage and wing fuel tanks
  - (*v*) Two engines

Each vital part is covered by some of the above mentioned triangular faces of the structure through which it can get lethal hits. Such triangles have been used to estimate kill probability of that vital component. Here pilot 1 and pilot 2 are nonredundant but fuel tanks and engines has been taken as redundant parts.

– *Data of air defence gun*

An Air Defence twin barrel gun with DA/VT fused ammunition is considered. Data of a typical air defence gun is given in Table 6.5 and critical and uncritical energy required to kill vital parts of the present aircraft has been given in Table 6.4.

Table 6.5: Gun data

Parameters	***Value***
System error	3 m rad
Number of barrels	2
Firing rate	5 rounds/gun barrel
Probability of DA fuze functioning	0.99
Probability of proximity fuze functioning	see Fig.7.3
Time of continuous firing of gun	3 seconds
Maximum range of gun	5000m
Minimum range of gun	500m
Maximum detecting range	10000m

Results of the model are computed and are shown in Table 6.6 for DA fused ammunition and in Table 6.7 for PF fused ammunition. First column gives cumulative kill probability of aircraft and other columns give kill probabilities of individual components. Similarly in Fig. 6.9, variation of kill probability of an aircraft vs number of rounds fired is shown when aircraft is engaged at range two kilo meters.

Table 6.6: Variation of probability of kill of aircraft and its five vital parts due to DA-fused ammunition (Engagement range = 2000m)

Rounds	***Aircraft***	***Avionics***	***Pilot 1***	***Pilot 2***	***Fuselage fuel tank***	***Wing fuel tank***	***Engine1***	***Engine 2***
2	0.0122	0.0008	.0100	0.0015	0.0009	0.0152	0.0010	0.00100
4	0.0247	0.0016	0.0200	0.0030	0.0018	0.0306	0.0021	0.0021
6	0.0373	0.0025	0.0305	0.0045	0.0028	0.0462	0.0032	0.0032
8	0.0503	0.0033	0.0410	0.0061	0.0039	0.0621	0.0044	0.0044
10	0.0634	0.0042	0.0517	0.0077	0.0049	0.0782	0.0056	0.0056
12	0.0769	0.0052	0.0627	0.0094	0.0060	0.0945	0.0069	0.0069
14	0.0905	0.0061	0.0738	0.0111	0.0072	0.1109	0.0082	0.0082

Contd...

16	0.1045	0.0071	0.0852	0.0129	0.0087	0.1275	0.0096	0.0096
18	0.1188	0.0081	0.0968	0.0148	0.0102	0.1443	0.0110	0.0110
20	0.1333	0.0092	0.1086	0.0166	0.0119	0.1614	0.0126	0.0126
22	0.1482	0.0103	0.1207	0.0186	0.0136	0.1787	0.0142	0.0142
24	0.1634	0.0115	0.1330	0.0206	0.0154	0.1964	0.0159	0.0159
26	0.1789	0.0127	0.1456	0.0227	0.0173	0.2143	0.0176	0.0177
28	0.1948	0.0139	0.1585	0.0249	0.0193	0.2325	0.0195	0.0195
30	0.2111	0.0152	0.1717	0.0271	0.0214	0.2510	0.0215	0.0215

It is seen that kill due to PF fused is much higher than that DA fused ammunition shell and increased with the number of rounds. Figure 6.10 gives variation of kill *vs*. Range. It is seen that kill decreases with higher opening range, for thirty rounds.

Figure 6.8 gives computer output of an aircraft generated by combining the triangular data from the drawings.

Table 6.7: Variation of probability of kill of the aircraft and its five vital parts due to VT-fused ammunition at range 2.0 km

Rounds	*Aircraft*	*Avionics*	*Pilot 1*	*Pilot 2*	*Fuselage fuel tank*	*Wing fuel tank*	*Engine 1*	*Engine 2*
2	0.0293	0.0193	0.0044	0.0037	0.0384	0.0441	0.0195	0.0196
4	0.0605	0.0380	0.0085	0.0071	0.0760	0.0870	0.0388	0.0389
6	0.0947	0.0564	0.0126	0.0105	0.1131	0.1290	0.0583	0.0585
8	0.1312	0.0747	0.0167	0.0141	0.1498	0.1701	0.0780	0.0783
10	0.1697	0.0928	0.0210	0.0177	0.1858	0.2103	0.0980	0.0983
12	0.2097	0.1106	0.0253	0.0214	0.2213	0.2495	0.1181	0.1185
14	0.2509	0.1281	0.029	0.0252	0.2562	0.2877	0.1384	0.1389
16	0.2928	0.1454	0.0343	0.0292	0.2904	0.3249	0.1588	0.1595
18	0.3350	0.1624	0.0389	0.0332	0.3239	0.3610	0.1795	0.1802
20	0.3772	0.1791	0.0436	0.0373	0.3567	0.3961	0.2003	0.2011
22	0.4191	0.1956	0.0484	0.0416	0.3889	0.4302	0.2212	0.2221
24	0.4604	0.2117	0.0533	0.0460	0.4203	0.4632	0.2424	0.2434
26	0.5010	0.2275	0.0583	0.0505	0.4510	0.4951	0.2637	0.2648
28	0.5404	0.2430	0.0634	0.0552	0.4809	0.5260	0.2851	0.2863
30	0.5785	0.2581	0.0686	0.0600	0.5100	0.5557	0.3067	0.3080

Fig. 6.8: *A three-dimensional computer output of the aircraft.*

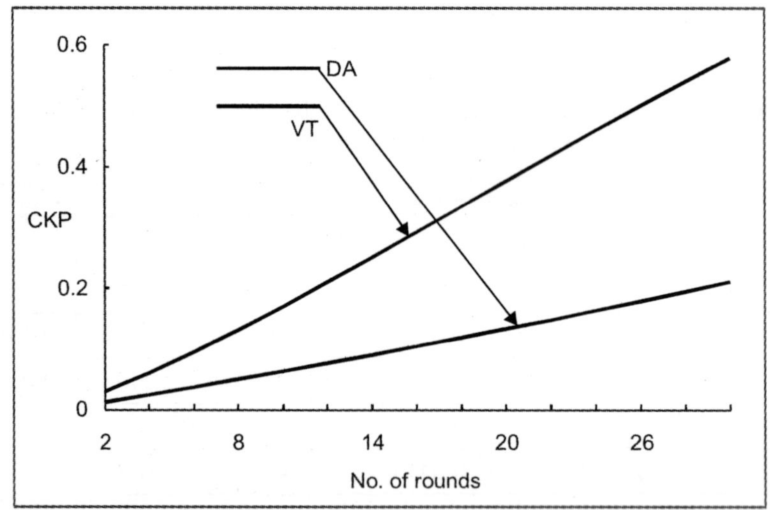

Fig. 6.9: *Variation of CKP vs. number of rounds (Engagement range 2000m).*

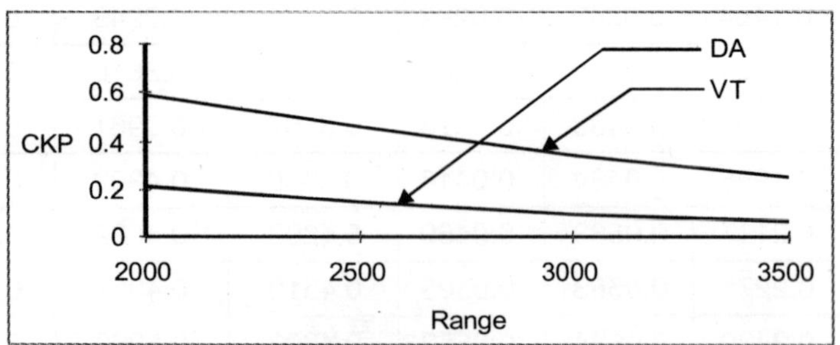

Fig. 6.10: *Variation of CKP vs range (30 rounds).*

APPENDIX 6.1

Let (x_0, y_0, z_0) be the centre of the N-plane forming a right handed system of axis, S-axis, T-axis and OG axis as shown in Fig. 6.2. The line GO is perpendicular to the ST plane with direction cosines $(-l_0, -m_0, -n_0)$ where

$$l_0 = \cos A \cdot \cos E = \frac{x_0}{|GO|}$$

$$m_0 = \sin A \cdot \cos E = \frac{y_0}{|GO|} \qquad \text{...(B.1)}$$

$$n_0 = \sin E = \frac{x_0}{|GO|}$$

S-axis which will lie in the so-called azimuthal plane will be normal to the elevation plane i.e., normal to the plane GOO' where O' is the projection of point O on GXY plane.

Equation of the plane GOO' is

$$l'_s x + m'_s y + n'_s z = 0 \qquad \text{...(B.2)}$$

Since this plane passes through the three points (O, O, O), (x_0, y_0, z_0) and (x_0, y_0, O) have,

$$x_0 l'_s + y_0 m'_s = 0 \qquad \text{...(B.3)}$$

$$x_0 l'_s + y_0 m'_s + z_0 n'_s = 0 \qquad \text{...(B.4)}$$

Solving (B.3), and (B.4) along with

$$l'^2_s + m'^2_s + n'^2_s = 1 \qquad \text{...(B.5)}$$

one gets the direction cosines of OS-axis as

$$\left(\frac{m_0}{\sqrt{1-n_0^2}}, \frac{-l_0}{\sqrt{1-n_0^2}}, 0\right) \qquad \text{...(B.6)}$$

Similarly for (l_t, m_t, n_t), we have following set of equations

$$l_t x_0 + m_t y_0 + n_t z_0 = 0 \quad \text{...(B.7)}$$

$$l'_s l_t + m'_s m_t + n'_s n_t = 0 \quad \text{...(B.8)}$$

$$l_t^2 + m_t^2 + n_t^2 = 1 \quad \text{...(B.9)}$$

Solving equations (B.7), (B.8) and (B.9) for l_t, m_t, n_t we get the directions cosines of OT axis as

$$\left(\frac{-n_0 l_0}{\sqrt{1-n_0^2}}, -\frac{m_0 n_0}{\sqrt{1-n_0^2}}, \sqrt{1-n_0^2} \right). \quad \text{...(B.10)}$$

APPENDIX 6.2

PROJECTION OF TRIANGULAR FACES OVER N-PLANE AND THEIR OVERLAPPING BY EACH OTHER

The projection of a triangular face over a plane normal to the line joining the centre of aircraft and the gun position (*N*-plane, in reference [66] it is called *D*-plane) will be a triangle. The vertices of the projected triangles on *N*-plane can be obtained from the equation (9) of ref [66].

For the determination of hit probability of a triangle or solid angle subtended, it is important to know, whether a particular triangular element is on the side of aircraft facing the source point of the projectile, or is on the other sides. This can be decided by considering the angle between the line joining the source point to the geometric centre of the triangular element and the normal to the triangular element at its geometric centre; as given below.

In the following paragraphs "source point of projectile" means the gun point while finding hit probabilities. But while estimation of solid angle the source point means the point where shell explodes i.e., source point of fragments.

Let (u_i, v_i, w_i), i = 1,3 are the co-ordinates of the vertices of a triangular face and $G(u_g, v_g, w_g)$ be the co-ordinates of the source point of the projectile w_0. r_0 t_0. frame-II. Let the $C(u_c, v_c, w_c)$ be the geometric centre of the triangular face 0, then

$$u_c = \frac{1}{3}\sum u_i$$

$$v_c = \frac{1}{3}\sum v_i$$

$$w_c = \frac{1}{3}\sum w_i$$

The *DCs*' of the line joining the points *G* and *C* are

$$l = (u_c - u_g)/D_s$$

$$m = (v_c - v_g)/D_s$$

$$n = (w_c - w_g)/D_s$$

$$D_s = \sqrt{\Sigma(u_c - u_g)^2}$$

DCs' (a, b, c) of the normal to a triangular surface co-ordinates of whose corners are (u_i, v_i, w_i), $i = 1, 2, 3$ can be obtained as

$$A = \begin{vmatrix} v_1 & w_1 & 1 \\ v_2 & w_2 & 1 \\ v_3 & w_3 & 1 \end{vmatrix}$$

$$B = \begin{vmatrix} u_1 & w_1 & 1 \\ u_2 & w_2 & 1 \\ u_3 & w_3 & 1 \end{vmatrix}$$

$$C = \begin{vmatrix} u_1 & v_1 & 1 \\ u_2 & v_2 & 1 \\ u_3 & v_3 & 1 \end{vmatrix}$$

$$a = \frac{A}{\sqrt{A^2 + B^2 + C^2}}$$

$$b = \frac{B}{\sqrt{A^2 + B^2 + C^2}}$$

$$c = \frac{C}{\sqrt{A^2 + B^2 + C^2}}$$

Let α is the angle between line $G\vec{C}$ and normal to the triangular face then

$$\alpha = \cos^{-1}(al + bm + cn)$$

$\alpha \le 90° \Rightarrow$ triangle is facing opposite to the source point G.

$\alpha > 90° \Rightarrow$ triangle is facing towards the source point G and can get impact provided if it is not being overlapped by some other Triangular-face of the aircraft.

To know that the i-th triangle is overlapped wholly or partially by another j-th triangle the following method is to be adopted.

Let (s_{ik}, t_{ik}), $k = 1, 3$ are the co-ordinates of the corners of projection of i-th triangle of the aircraft over N-plane with respect to point G and (s_{jk}, t_{jk}), $k = 1, 3$ are the co-ordinates of the projection of j-th triangle over N-plane. These triangles are further subdivided into smaller rectangular meshes, in order to assess the overlapped area.

Let (s_0, t_0) be the central point of a typical rectangle of the i-th triangle. If this point falls on in the j-th triangle formed by the vertices (s_{jk}, t_{jk}), $k = 1, 3$ then it implies that this rectangle is overlapped by the j-th triangular face. If it is not covered by j-th triangle, other triangles are tested. Similar test is applied to all the rectangular elements of the i-th triangle. Thus if all the rectangular elements are not covered by any other triangle, it implies that this triangle is not being overlapped by any of the triangles and can be considered to find solid angle or hit probability. Let a rectangle with centre $c_p(s_0, t_0)$ is overlapped by j-th triangle. In that case we have to check whether the j-th triangular face is near to the source point of the projectile or the i-th triangular face is nearer. Whichever triangle is nearer will be overlapping other. It can be done in the following steps.

(*i*) Let us define a III-co-ordinate system, *OST*, with origin at 0 and s-t plane being normal to line joining projectile and centre of the aircraft.

(*ii*) Let direction cosines of the line, joining points $c_p(s_0, t_0)$ and source point of the projectile G with respect to frame-III are (l'_3, m'_3, n'_3)

(*iii*) Let (l'_1, m'_1, n'_1) be the direction cosines of $\vec{G}C_p$ with reference to frame-I. Thus

$$l'_1 = l'_3 l_s + m'_3 l_t + n'_3 l_r$$

$$m'_1 = l'_3 m_s + m'_3 m_t + n'_3 m_r$$

$$n'_1 = l'_3 n_s + m'_3 n_t + n'_3 n_r$$

where (l_s, m_s, n_s); (l_t, m_t, n_t) and (l_r, m_r, n_r) are the direction cosines of the axes of frame-III with respect to frame-I.

Thus and the DC's of the line $\vec{G}C_p$ with respect to frame-II say (l'_2, m'_2, n'_2)

$$l'_2 = l'_1 l_1 + m'_1 m_1 + n'_1 n_1$$

$$m'_2 = l'_1 l_2 + m'_1 m_2 + n'_1 n_2$$

$$n'_2 = l'_1 l_3 + m'_1 m_3 + n'_1 n_3$$

(*iv*) Finally find the equation of the line with respect to frame-II as the line passes though some point whose co-ordinate with respect to frame-II are known.

(*v*) Let the line meet the i-th triangular face at point O_i with co-ordinates $o_i(uo_i, vo_i, wo_i)$

$$uo_i = u_g + l'_2 R_0$$

$$uo_i = v_g + m'_2 R_0$$

$$uo_i = w_g + n'_2 R_0$$

$$R_0 = -\left(\frac{au_g + bv_g + cw_g + d_i}{al_2' + bm_2' + cn_2'}\right)$$

$$d_i = -au_1 - bv_1 - cw_1$$

(*vi*) Find the intersection of the line with *j*-th triangular face $Uj_i(UO_j, VO_j, WO_j)$ as explained above.

(*vii*) Find the distances of the line GO_i *and* GO_j If $GO_i > GO_j$ implies that this rectangle is being overlapped by *j*-th Δ face and need not be considered to find hit probability or solid angle i.e., $GO_i < GO_j$ meant that the rectangle is not being overlapped by *j*-th triangular face.

(*viii*) Same methodology can be used to check overlapping by other triangular faces i.e., for all *js*'.

(*ix*) The same method is to be repeated for all the rectangles of the *i*-th triangle on *N*-plane.

SIMULATION OF QUEUING SYSTEMS

When we enter a bank, especially on Saturday, long queues on a counter are found, and one has to wait for hours, if it is a public bank. Reason for long queues may be due to less number of counters in the banks. But on the other hand if bank opens more number of counters, then on normal days, when customers are less in numbers, counter remains idle. Whether it is a bank, or a theater or waiting for a bus, we find queues everywhere in our day to day life. Theory of queuing is to sort out such problems. *Agner Krarup Erlang*[1], a Danish engineer who worked for the Copenhagen Telephone Exchange, published the first paper on queuing theory in 1909. Application of queuing theory to machine shop, where jobs arrive in queues, and wait for completion, is another example of queues. In this chapter, attempt will be

1. A.K. Erlang was the first person to study the problem of telephone networks. By studying a village telephone exchange he worked out a formula, now known as Erlang's formula, to calculate the fraction of callers attempting to call someone outside the village that must wait because all of the lines are in use. Although Erlang's model is a simple one, the mathematics underlying today's complex telephone networks is still based on his work.

He was born at Lønborg, in Jutland, Denmark. His father, Hans Nielsen Erlang, was the village schoolmaster and parish clerk. His mother was Magdalene Krarup from an ecclesiastical family and had a well known Danish mathematician, Thomas Fincke, amongst her ancestors. He had a brother, Frederik, who was two years older and two younger sisters, Marie and Ingeborg. Agner spent his early school days with them at his father's schoolhouse. Evenings were often spent reading a book with Frederik, who would read it in the conventional way and Agner would sit on the opposite side and read it upside down. At this time one of his favourite subjects was astronomy and he liked to write poems on astronomical subjects. When he had finished his elementary education at the school he was given further private tuition and succeeded in passing the Praeliminaereksamen (an examination held at the University of Copenhagen) with distinction. He was then only 14 years old and had to be given special entrance permission. Agner returned home where he remained for two years, teaching at his father's school for two years and continuing with his studies. He also learnt French and Latin during this period. By the time he was 16 his father wanted him to go to university but money was scarce. A distant family relation provided free accommodation for him while he prepared

made to model science of queues. The basic concept of queuing theory is the optimization of wait time, queue length, and the service available to those standing in a queue. Cost is one of the important factors in the queuing problem. Waiting in queues incur cost, whether human are waiting for services or machines waiting in a machine shop. On the other hand if service counter is waiting for customers that also involves cost. In order to reduce queue length, extra service centers are to be provided but for extra service centers, cost of service becomes higher. On the other hand excessive wait time in queues is a loss of customer time and hence loss of customer to the service station. Ideal condition in any service center is that there should not be any queue. But on the other hand service counter should also be not idle for long time. Optimization of queue length and wait time is the object theory of queuing.

Let us see how this situation is modeled. First step is to know the arrival time and arrival pattern of customer. Here customer means an entity waiting in the queue. One must know from the past history, the time between the successive arrival of customers or in the case of machine shop, the job scheduling. Also arrival of number of customers vary from day to day. On Saturdays, number of customers may be more than that on other days. What is the probability that a customer will arrive in a given span of time, is important to know.

In order to maximize the profit, the major problem faced by any management responsible for a system is, how to balance the cost associated with the waiting, against the cost associated with prevention of waiting. An analysis of queuing system will provide answers to all these questions. However, before looking at how queuing problem is to be solved, the general framework of a queuing system should be understood.

A queuing system involves customers arriving at a constant or variable time rate for service at a service station. Customers can be students waiting for registration in college, aeroplane queuing for landing at airfield, or jobs waiting in machines shop. If the customer after arriving, can enter the service center, good, otherwise they have to wait for the service and form a queue. They remain in queue till they are provided the service. Sometimes queue being too long, they will leave the queue and go, resulting a loss of customer. Customers are to be serviced at a constant or variable rate before they leave the service station. A typical queuing system is shown in Fig. 7.1.

---

for his University entrance examinations at the Frederiksborg Grammar School. He won a scholarship to the University of Copenhagen and completed his studies there in 1901 as an M. A. with mathematics as the main subject and Astronomy, Physics and Chemistry as secondary subjects.

Over the next 7 years he taught in various schools. Even though his natural inclination was towards scientific research, he proved to have excellent teaching qualities. He was not highly sociable, he preferred to be an observer, and had a concise style of speech. His friends nicknamed him "The Private Person". He used his summer holidays to travel abroad to France, Sweden, Germany and Great Britain, visiting art galleries and libraries. While teaching, he kept up his studies in mathematics and natural sciences. He was a member of the Danish Mathematicians' Association through which he made contact with other mathematicians including members of the Copenhagen Telephone Company. He went to work for this company in 1908 as scientific collaborator and later as head of its laboratory.

Erlang at once started to work on applying the theory of probabilities to problems of telephone traffic and in 1909 published his first work on it "The Theory of Probabilities and Telephone Conversations"[1] proving that telephone calls distributed at random follow Poisson's law of distribution. At the beginning he had no laboratory staff to help him, so he had to carry out all the measurements of stray currents. He was often to be seen in the streets of Copenhagen, accompanied by a workman carrying a ladder, which was used to climb down into manholes.

7.0 Symbols Used

Unless and otherwise stated, the following standard terminology and notations will be used in this chapter.

State of system = number of customers in the queuing system (queue and server).

Queue length = number of customers waiting for service to begin.

= (state of system) – (number of customers being served).

$N(t)$ = number of customers in the queuing system at time t ($t \geq 0$)

$P_n(t)$ = probability of exactly n customers in the queuing system at time t, given number at time $t = 0$.

s = number of servers (parallel service channels) in queuing system.

λ_n = mean arrival rate (expected number of arrivals per unit time) of new customers when n customers are in system.

μ_n = mean service rate for overall system (expected number of customers completing service per unit time) when n customers are in the system.

When λ_n is a constant for all n, is denoted by λ.

When the mean service rate per busy server is a constant for all $n \geq 1$, this constant is denoted by μ (single server).

$\mu_n = s\mu$ when $n \geq s$, that is, when s servers are busy.

The queuing system is classified in general as follows.

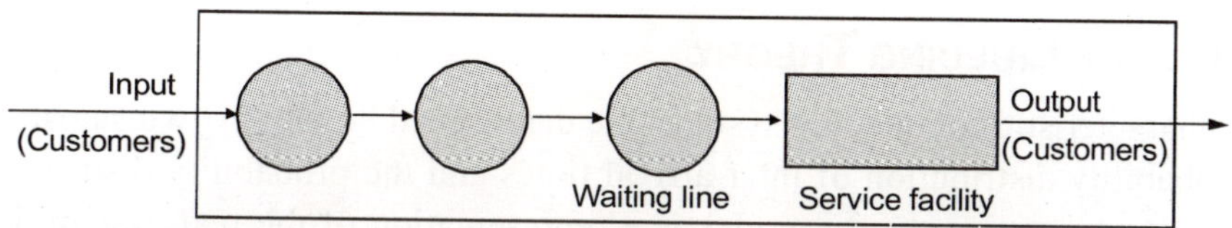

Fig. 7.1: *Single queue-single server queuing system.*

Because of the growing interest in his work several of his papers were translated into English, French and German. He wrote up his work in a very brief style, sometimes omitting the proofs, which made the work difficult for non-specialists in this field to understand. It is known that a researcher from the Bell Telephone Laboratories in the USA learnt Danish in order to be able to read Erlang's papers in the original language.

His work on the theory of telephone traffic won him international recognition. His formula for the probability of loss was accepted by the British Post Office as the basis for calculating circuit facilities. He was an associate of the British Institution of Electrical Engineers.

Erlang devoted all his time and energy to his work and studies. He never married and often worked late into the night. He collected a large library of books mainly on mathematics, astronomy and physics, but he was also interested in history, philosophy and poetry. Friends found him to be a good and generous source of information on many topics. He was known to be a charitable man, needy people often came to him at the laboratory for help, which he would usually give them in an unobtrusive way. Erlang worked for the Copenhagen Telephone Company for almost 20 years, and never having had time off for illness, went into hospital for an abdominal operation in January 1929. He died some days later on Sunday, 3rd February 1929.

Interest in his work continued after his death and by 1944 "Erlang" was used in Scandinavian countries to denote the unit of telephone traffic. International recognition followed at the end of World War.

1. **Calling source.** or the population from which customers are drawn. Calling source may be finite or infinite. When queue is so long that arrival of one more customer does not effect the queue length, we call it *infinite source* of customers. A reverse of this situation, when queue length is not long and incoming or outgoing of one-customer affects the queue; we call it a *finite source* of customers.
2. **The input or arrival process.** This includes the distribution of number of arrivals per unit of time, the number of queues that are permitted to be formed, the maximum queue length, and the maximum number of customers desiring service.
3. **The service process.** This includes time allotted to serve a customer, number of servers and arrangement of servers. Simplest case is single queue and single server.

7.1 Kendall's Notation

We will be frequently using notation for queuing system, called Kendall's notation, that is, *V/W/X/Y/Z*, where, *V*, *W*, *X*, *Y*, *Z* respectively indicate arrival pattern, service pattern, number of servers, system capacity, and queue discipline. The symbols used for the probability distribution for inter arrival time, and service time are, *D* for deterministic, *M* for exponential and *Ek* for Erlang. Similarly FIFO (First in First out), LIFO (Last in First out), etc., for queue discipline.

If the capacity *Y* is not specified, it is taken as infinity, and if queue discipline is not specified, it is FIFO (First in First Out). For example *M/D/2/5/*FIFO stands for a queuing system having exponential arrival times, deterministic service time, 2 servers, capacity of 5 customers, and first in first out discipline. If notation is given as *M/D/*2 means exponential arrival time, deterministic service time, 2 servers, infinite service capacity, and FIFO queue discipline.

7.2 Principle of Queuing Theory

The operating characteristics of queuing systems are determined largely by two statistical properties, namely, the probability distribution of inter arrival times and the probability distribution of service times. To formulate a queuing theory model as a representation of the real system, it is necessary to specify the assumed form of each of these distributions. We will now try to model this situation, under given conditions. For the case of simplicity, we will assume for the time being, that there is single queue and only one server serving the customers. We make the following assumptions.

- **First-in, First-out (FIFO):** Service is provided on the first come, first served basis.
- **Random:** Arrivals of customers is completely random but at a certain arrival rate.
- **Steady state:** The queuing system is at a steady state condition.

The above conditions are very ideal conditions for any queuing system and assumptions are made to model the situation mathematically. First condition only means irrespective of customer, one who comes first is attended first and no priority is given to anyone. Second conditions says that arrival of a customer is random and is expected anytime after the elapse of first mean time of interval (τ say). In a given interval of time (called mean time of arrival τ, between two customers) only one customer is expected to come. This is equivalent to saying that the number of arrivals per unit time is a random variable with a Poisson's distribution. This distribution is used when chances of occurrence of an event out of a large sample is small. That is, if

X = number of arrivals per unit time, then, probability distribution function of arrival is given as,

$$f(x) = \Pr(X = x) = \frac{e^{-\lambda}\lambda^x}{x!}, \begin{cases} x = 0, 1, 2, ... \\ \lambda > 0 \end{cases} \quad ...(7.1)$$

$$E(X) = \lambda$$

where λ is the average number of arrivals per unit time ($1/\tau$), and x is the number of customers per unit time. This pattern of arrival is called **Poisson's arrival pattern**.

It is interesting to know that second assumption leads us to the result that inter arrival time T follows an exponential distribution[1], with the same mean parameter λ. To prove this, let us assume T = time between consecutive arrivals. If an arrival has already occurred at time t = 0, the time to the next arrival is less than t, if and only if there are more than one arrival in time interval [0, t]. This probability $G(t)$ that inter-arrival time is less than t can be defined as,

$$G(t) = \Pr(T < t) = \sum_{x=1}^{\infty} e^{-\lambda t}(\lambda t)^x / x!$$

where x is the number of arrivals in time t. But

$$\sum_{x=0}^{\infty} e^{-\lambda t}(\lambda t)^x / x! = 1$$

Therefore

$$\sum_{x=1}^{\infty} e^{-\lambda t}(\lambda t)^x / x! = 1 - e^{-\lambda t}$$

$$G(t) = \Pr(T < t) = 1 - e^{-\lambda t}$$

---

1. An Alternative proof

The probability of arrival of a customer during a very small time interval Δt is $\frac{\Delta t}{\tau}$. Hence probability of a customer not arriving during time Δt is $(1 - \frac{\Delta t}{\tau})$. Now if

$h(t)$ = probability that the next customer does not arrive during the interval t given that the previous customer arrived at time t = 0, and likewise.

$h(t + \Delta t)$ = probability that the next customer does not arrive during the interval $(t + \Delta t)$ given that the previous customer arrived at time t = 0.

Since the arrival of customers in different periods are independent events (i.e., the queue has no memory), we write

$$h(t + \Delta t) = h(t) \cdot (1 - \frac{\Delta t}{\tau}) \quad ...(7.1a)$$

or

$$\frac{h(t + \Delta t) - h(t)}{\Delta t} = -\frac{h(t)}{\tau}$$

First equation of (7.1*a*) means, probability that customer does not arrive in interval $(t + \Delta t)$ is equal to the probability that he does not arrive in the interval t and also in the interval Δt.

Contd...

Since $G(t)$ is the cumulative distribution of T, the density distribution of T is given by

$$g(t) = \frac{d[G(t)]}{dt} = \frac{d(1-e^{-\lambda t})}{dt} = \lambda e^{-\lambda t} \quad ...(7.2)$$

Equation (7.2) is the exponential probability density function discussed earlier in chapter two. Figure 7.2 gives plot of exponential density function and its cumulative distribution function.

The curve in Fig. 7.2(*b*) gives the probability that the next customer arrives by the time *t*, given that the preceding customer arrived at time zero. It is appropriate to mention here that inverse of inter arrival time τ is denoted by λ and is the average number of customers at the server per unit time.

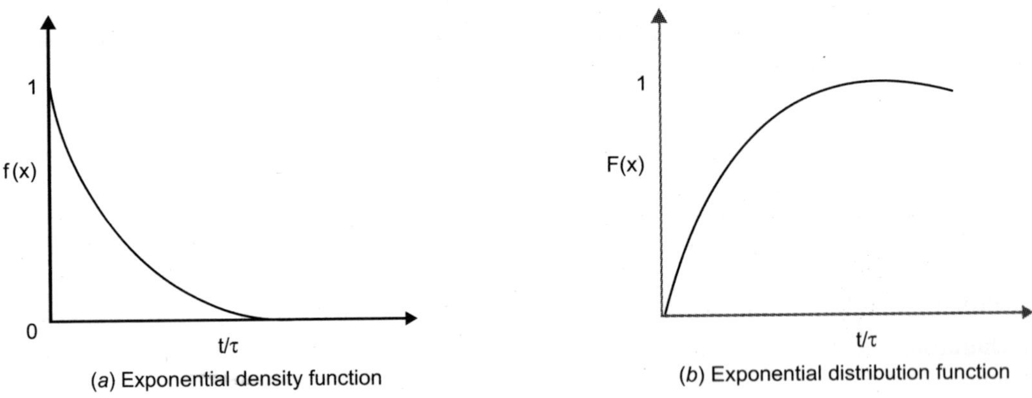

(*a*) Exponential density function (*b*) Exponential distribution function

Fig. 7.2: *Inter arrival time of customers.*

Taking limits on both sides as Δt tends to zero, one gets

$$\frac{d}{dt}[h(t)] = -\frac{h(t)}{\tau}$$

Integral of this equation is

$$h(t) = ce^{-t/\tau} \quad ...(7.2a)$$

Since it was assumed that at time $t = 0$, a customer had just arrived, therefore the probability of non-arrival at time $t = 0$ is one. That is $h(0) = 1$, and therefore constant of integration in (7.2*a*) is unity.

The relation $h(t) = e^{-t/\tau}$ is derived with two very simple assumptions, that is, (*i*) constancy of a long-term average and (*ii*) statistical independence of arrivals. Thus equation (7.2*a*) with $c = 1$ gives the probability that the next customer does not arrive before time *t* has elapsed since the arrival of the last customer.

The probability that a customer arrives during infinitesimal interval between *t* and $t + \Delta t$ is given as the product of (*i*) the probability that no customer arrives before the time *t* and (*ii*) the probability that exactly one customer arrives during the time Δt. That is $\left(e^{-t/\tau}\right)\cdot\left(\frac{\Delta t}{\tau}\right)$

In other words, the probability density function of the inter arrival time is

$$g(t) = \frac{1}{\tau}\left(e^{-t/\tau}\right) \quad ...(7.3a)$$

which is the distribution for inter arrival time.

We give below few examples to understand these results.

Example 7.1: In a single pump service station, vehicles arrive for fueling with an average of 5 minutes between arrivals. If an hour is taken as unit of time, cars arrive according to Poison's process with an average of λ = 12 cars/hr. The distribution of the number of arrivals per hour is,

$$f(x) = \Pr(X = x) = \frac{e^{-\lambda}\lambda^x}{x!} = \frac{e^{-12}12^x}{x!}, \begin{cases} x = 0, 1, 2, ... \\ \lambda > 0 \end{cases}$$

$$E(X) = 12 \text{ cars/hr}$$

The distribution of the time between consecutive arrivals T is,

$$g(t) = 12e^{-12t};\ t > 0$$

$$E(T) = \frac{1}{12} \text{ hr between arrivals.}$$

Third assumption (steady state) means queuing system has been operating long enough to be independent of the initial state of the system and is independent of time. That is, the system has reached a state of equilibrium with respect to time. The distribution of the number of arrivals per unit time and the distribution of the service time do not change with time.

7.3 Arrival of k Customers at Server

In the present section concept of arrival of single customer in time t will be extended to arrival of k customers in time t. It has been seen that time for single customer arrival follows an exponential distribution.

Let $q_k(t)$ be the probability that k, (here k = 0, 1, 2, 3,...), customers arrive in time t when at t = 0, no customer arrived at the server. Then probability that single customer arrives between time $(t + \Delta t)$is given by $q_1(t + \Delta t)$, which is given by,

$q_1(t+\Delta t)$ = (probability that no arrival takes place between time zero and t) $\cdot$ (probability that single arrival takes place during time Δt) + (probability that single arrival takes place between time zero and t) $\cdot$ (probability that no arrival takes place during time Δt).

$$= f(t)\cdot\frac{\Delta t}{\tau} + q_1(t)\cdot(1 - \frac{\Delta t}{\tau})$$

where $f(t) = e^{-t/\tau}$. Here $f(t)$ is a Poisson's distribution function for x = 0, thus

$$\frac{q_1(t+\Delta t) - q_1(t)}{\Delta t} = \frac{1}{\tau}[f(t) - q_1(t)]$$

When limit $\Delta t \to 0$, this equation becomes,

$$\frac{dq_1}{dt} = \frac{1}{\tau}[f(t) - q_1(t)]$$

Solution of this differential equation is,

$$q_1(t) = \frac{t}{\tau}e^{-t/\tau} = \frac{t}{\tau}f(t)$$

Now we extend the above logic for two customers in the queue.

$q_2(t+\Delta t)$ = (probability that one arrival takes place between time zero and t). (probability that single arrival takes place during time Δt) + (probability that two arrivals take place between time zero and t). (probability that no arrival takes place during time Δt).

$$= q_1(t)\cdot\left(\frac{\Delta t}{\tau}\right) + q_2(t)\left(1-\frac{\Delta t}{\tau}\right)$$

$$\frac{q_2(t+\Delta t)-q_2(t)}{\Delta t} = \frac{1}{\tau}[q_1(t)-q_2(t)]$$

When limit $\Delta t \to 0$, this equation becomes,

$$\frac{dq_2}{dt} = \frac{1}{\tau}[q_1(t)-q_2(t)]$$

Above equation can be integrated as,

$$q_2(t) = \frac{1}{2!}\left(\frac{t}{\tau}\right)^2 f(t)$$

We generalize this logic for k customers to arrive between time zero and t as,

$$q_k(t) = \frac{1}{k!}\left(\frac{t}{\tau}\right)^k f(t) \qquad \text{...(7.4)}$$

where $f(t) = e^{-t/\tau}$.

Expression (7.4) is known as *Poisson Distribution Formula*, which was assumed in equation (7.1) as an arrival pattern for a unit time ($t = 1$). It is the most important and widely used distribution and has been discussed earlier in chapter two.

It is seen that, if the arrival time is distributed exponentially, the number of arrivals are given by Poisson's distribution and vice versa. It is to be emphasized here that Poisson's method of arrival is just one of the arrival pattern in queuing theory, which results from the three assumptions, that is

- Successive arrivals are statistically independent of each other
- There is a long term inter arrival constant τ and
- The probability of an arrival taking place during a time interval Δt is directly proportional to Δt.

7.3.1 Exponential Service Time

Let us make the similar assumptions about the servicing process too, namely

1. The statistical independence of successive servicing
2. The long time constancy of service time and
3. Probability of completing the service for a customer during a time interval Δt is proportional to Δt.

Therefore, as in the case of inter arrival time, we get

$$g(t) = e^{-t/v} \quad ...(7.5)$$

where $g(t)$ is the probability that a customer's service could not be completed in time t, given that previous customer's service was completed at time zero, and ν is the long term average service time. Average number of customers served at the server per unit time are μ which is inverse of ν.

7.4 Queuing Arrival-service Model

So far we have discussed the arrival pattern of customers. Now we develop an algorithm giving arrival service pattern in a queue of n customers. Following assumptions are made,

(*a*) Arrival to the system occurs completely random

(*b*) Arrivals form a single queue

(*c*) First in first out discipline (FIFO)

(*d*) Departure from the system occurs completely at random.

(*e*) The probability of an arrival in the interval t to $t + \Delta t$ at time t is $\lambda\Delta t$.

(*f*) The probability of a departure in the interval t to $t + \Delta t$ at time t is $\mu\Delta t$.

At any time t the probability of the service counter being busy is

$$\frac{\textit{average service time}}{\textit{average arival time}} = \frac{\nu}{\tau} = \frac{\lambda}{\mu} = \rho \quad ...(7.6)$$

where ρ is called the utilization factor of the service facility. This is also the average number of customers in the service facility. Thus probability of finding service counter free is

$$(1 - \rho) \quad ...(7.6a)$$

That is there is zero customer in the service facility. Let $P_n(t)$ be the probability of exactly n customers being in the system at time t. Let $\Delta t > 0$ be a small interval of time. The probability of one customer arriving and no customer departing during the interval Δt is

$$\lambda \cdot \Delta t \cdot (1 - \mu \Delta t)$$

Similarly, probability of one customer arriving and one customer leaving during the interval Δt is

$$(\lambda \cdot \Delta t) \cdot (\mu \cdot \Delta t)$$

The probability of no customer arriving and one customer leaving is

$$(1 - \lambda \cdot \Delta t) \cdot (\mu \cdot \Delta t)$$

The probability of no customer arriving and no customer leaving is

$$(1-\lambda\cdot\Delta t)\cdot(1-\mu\cdot\Delta t)$$

It is assumed that time interval Δt is so small that no more than one arrival and one departure can take place in this interval. These are the only possibilities that could occur during this interval.

For the queuing system to have n customers at time $(t + \Delta t)$, it must have either n or $(n + 1)$ or $(n - 1)$ customers at time t. The probability that there are n customers in the system at time $(t + \Delta t)$, can therefore be expressed as the sum of these three possibilities. Thus for any $n > 0$ we can write,

$$\begin{aligned} P_n(t+\Delta t) &= P_n(t)\cdot(\lambda\Delta t)\cdot(\mu\Delta t)+P_n(t)\cdot(1-\lambda\Delta t)\cdot(1-\mu\Delta t) \\ &\quad +P_{n+1}(t)\cdot(1-\lambda\Delta t)\cdot(\mu\Delta t) \\ &\quad +P_{n-1}(t)\cdot(\lambda\Delta t)\cdot(1-\mu\Delta t) \end{aligned}$$

From the above equation one gets,

$$\begin{aligned} \frac{P_n(t+\Delta t)-P_n(t)}{\Delta t} &= P_n(t)\cdot(\lambda\mu\Delta t) + P_n(t)[-\lambda-\mu+\lambda\cdot\mu\Delta t) \\ &\quad +P_{n+1}(t)\cdot(1-\lambda\Delta t)\cdot(\mu) \\ &\quad +P_{n-1}(t)\cdot(\lambda)\cdot(1-\mu\Delta t) \\ &= \mu P_{n+1}(t)+\lambda P_{n-1}(t)-P_n(t)(\lambda+\mu) \\ &\quad +\lambda\cdot\mu\Delta t[2P_n(t)-P_{n+1}(t)-P_{n-1}(t)] \end{aligned}$$

Taking limits of both sides of this equation when Δt tends to zero,

$$\frac{dP_n(t)}{dt} = \mu P_{n+1}(t)+\lambda P_{n-1}(t)-(\lambda+\mu)\cdot P_n(t) \qquad ...(7.7)$$

This equation holds for all $n > 0$. When $n = 0$, the contributions made by the term P_{n-1} would be zero. Therefore

$$\frac{dP_0(t)}{dt} = \mu P_1(t)-\lambda\cdot P_0(t) \qquad ...(7.8)$$

If $\rho < 1$, after the passage of sufficiently long time the queue would reach an equilibrium, and $P_n(t)$ would converge to a constant. Thus its derivative can be put equal to zero in equilibrium condition. The equations (7.7) and (7.8) become,

$$P_{n+1}(t)+\frac{\lambda}{\mu}P_{n-1}(t)-\left(\frac{\lambda}{\mu}+1\right)\cdot P_n(t)=0$$

$$P_{n+1}(t) = (1+\rho)\cdot P_n(t)-\rho P_{n-1}(t), \text{ for } n\geq 1 \qquad ...(7.9)$$

$$P_1(t) = \frac{\lambda}{\mu}\cdot P_0(t)=\rho P_0(t) \qquad ...(7.10)$$

Using equation (7.10) in equation (7.9) repeatedly we get,

$$P_n = \rho^n P_0 \text{ for all } n > 0 \text{ and } \rho < 1. \qquad ...(7.11)$$

Now since $\sum_{n=0}^{\infty} P_n = 1$, therefore

$$P_0 \sum_{n=0}^{\infty} \rho^n = 1$$

or

$$P_0 \frac{1}{1-\rho} = 1$$

since $\rho < 1$, therefore $P_0 = 1-\rho$

which means there is no body (zero person in the system) in the system (queue plus server) and service counter is free, which is the same result as in equation (7.6*a*).

We define average number of customers at time *t*, in the system as $\bar{L}_S$. Here bar over *L* depicts average length of queue and subscript *S* is for system. Similar terminology will be used for other symbols in the following sections. $\bar{L}_S$ is given by

$$\bar{L}_S = \sum_{n=0}^{\infty} nP_n = P_0 \sum n\rho^n = \frac{\rho}{1-\rho}$$

since

$$P_0 \sum n\rho^n = \rho(1-\rho)(1+2\rho+3\rho^2+4\rho^3+\cdots)$$

$$= \rho(1-\rho)(1-\rho)^{-2}$$

Thus

$$\bar{L}_S = \frac{\lambda}{\mu-\lambda} = \frac{\rho}{(1-\rho)} \qquad ...(7.12)$$

Probability of *n* customers being in the system can also be expressed as

$$P_n = \rho^n(1-\rho) \qquad ...(7.13)$$

Since $P_0 = (1-\rho)$. Similarly probability of *n* customers in the queue is same as the probability of $(n+1)$ customers in the system i.e.,

$$P_{n+1} = \rho^{n+1}(1-\rho), \text{ for } n > 0 \qquad ...(7.14)$$

Probability of more than *n* customers being in the system is

$$P(N > n) = 1 - \sum_{i=0}^{n} \rho^i (1-\rho)$$

$$= 1-[(1-\rho)+\rho(1-\rho)+\rho^2(1-\rho)+\cdots \qquad ...(7.15)$$

$$+ \rho^{n-1}(1-\rho)+\rho^n(1-\rho)]$$

$$= 1-[1-\rho^{n+1}] = \rho^{n+1}$$

These and similar other statistics about the queue are called the operating characteristics of the queuing system.

Average number of customers in the queue $\bar{L}_Q$ is same as expected number in the system – the expected number in the service facility i.e.,

$$\bar{L}_Q = \bar{L}_S - \rho = \frac{\lambda}{\mu - \lambda} - \frac{\lambda}{\mu} = \frac{\lambda^2}{\mu(\mu - \lambda)} = \frac{\rho^2}{(1-\rho)} \quad ...(7.16)$$

Average time a customer spends in the system is denoted by $\overline{W}_S$, and is equal to expected number of customers in the system at time t, divided by number of customers arrived in unit time i.e.,

$$\overline{W}_S = \frac{\lambda}{\mu - \lambda} \cdot \frac{1}{\lambda} = \frac{1}{(\mu - \lambda)} \quad ...(7.17)$$

Average time a customer spends in the queue $\left(\overline{W}_Q\right)$ is same as average time a customer spends in the system – average time a customer spends in the server i.e.,

$$\overline{W}_Q = \overline{W}_S - \frac{1}{\mu} = \frac{\lambda}{\mu(\mu - \lambda)}$$

= average time a customer spends in the queue. ...(7.18)

Similarly few other parameters can be defined as follows.

Now expected time T to serve a customer is given by,

$$E(T = t) = \frac{1}{\mu} = \upsilon$$

and time for which server remains idle in t seconds is given by $(1-\rho)t/\upsilon$, thus

Probability that the time in the system is greater than t is given by,

$$P(T > t) = e^{-\mu(1-\rho)t} \quad ...(7.19)$$

Similarly probability of more than k customers in the system is,

$$P(n > k) = \left(\frac{\lambda}{\mu}\right)^{k+1} \quad ...(7.20)$$

Below, we give few examples to illustrate these statistics.

Example 7.2: In a tool crib manned by a single assistant, operators arrive at the tool crib at the rate of 10 per hour. Each operator needs 3 minutes on the average to be served. Find out the loss of production due to waiting of an operator in a shift of 8 hours if the rate of production is 100 units per shift.

Solution: Arrival rate (λ) = 10 per hour

Service rate (μ) = 60/3 = 20 per hour

Average waiting time in the queue

$$\left(\overline{W}_Q\right) = \frac{\lambda}{\mu(\mu - \lambda)} = \frac{10}{20(20-10)} = \frac{1}{20} \text{ hour}$$

Average waiting time per shift of 8 hours = 8/20 = 2/5 hr.

$$\therefore \quad \text{loss of production due to waiting} = \frac{100}{8} \times \frac{2}{5} = 5 \text{ units}$$

Example 7.3: At the ticket counter of football stadium, people come in queue and purchase tickets. Arrival rate of customers is 1/min. It takes at the average 20 seconds to purchase the ticket.

(*a*) If a sport fan arrives 2 minutes before the game starts and if he takes exactly 1.5 minutes to reach the correct seat after he purchases a ticket, can the sport fan expects to be seated for the tip-off?

(*b*) What is the probability the sport fan will be seated for the start of the game?

(*c*) How early must the fan arrive in order to be 99% sure of being on the seat for the start of the game?

Solution: (*a*) A minute is used as unit of time. Since ticket is disbursed in 20 seconds, this means, three customers enter the stadium per minute, that is service rate is 3 per minute. Therefore,

$$\lambda = 1 \text{ arrival/min}$$

$$\mu = 3 \text{ arrivals/min}$$

$$\left(\overline{W_S}\right) = \text{waiting time in the system} = \frac{1}{\mu - \lambda} = \frac{1}{2}$$

The average time to get the ticket and the time to reach the correct seat is 2 minutes exactly, so the sports fan can expect to be seated for the tip-off.

(*b*) This is equivalent to the probability the fan can obtain a ticket in less than or equal to half minute.

$$P(T < 1/2) = 1 - P(T > 1/2)$$

$$P\left(T < \frac{1}{2}\right) = 1 - e^{-\mu(1-\frac{\lambda}{\mu})t}$$

$$= 1 - e^{-3(1-\frac{1}{3})\cdot(\frac{1}{2})}$$

$$= 1 - e^{-1} = 0.63$$

(*c*) For this problem, we need to determine *t* such that, probability of arrival of fan in the playground just before the start is 0.99, or probability that he does not arrive in time is 0.01.

$$P\left(T < \frac{1}{2}\right) = 1 - e^{-\mu(1-\frac{\lambda}{\mu})t}$$

$$\text{But} \qquad P(T > t) = e^{-\mu(1-\frac{\lambda}{\mu})t}$$

$$\Rightarrow \qquad = e^{-3(1-\frac{1}{3})t} = 0.01$$

$$\Rightarrow \qquad -2t = \ln(0.01)$$

$$\therefore \qquad t = 2.3 \text{ minutes}$$

Thus, the fan can be 99% sure of spending less than 2.3 minutes obtaining a ticket (awaiting for the purchase of ticket). Since it can take exactly 1.5 minutes to reach the correct seat after purchasing the ticket, the fan must arrive 3.8 minutes (2.3 + 1.5 = 3.8) early to be 99% of seeing the tip-off.

Example 7.4: Customers arrive in a bank according to a Poisson's process with mean inter arrival time of 10 minutes. Customers spend an average of 5 minutes on the single available counter, and leave. Discuss (*a*) What is the probability that a customer will not have to wait at the counter, (*b*) What is the expected number of customers in the bank, (*c*) How much time can a customer expect to spend in the bank.

Solution: We will take an hour as the unit of time. Thus

$$\lambda = 6 \text{ customers/hour,}$$

$$\mu = 12 \text{ customers/hour.}$$

(*a*) The customer will not have to wait if there are no customers in the bank. Thus

$$P_0 = 1 - \frac{\lambda}{\mu}$$

$$= 1 - 6/12 = 0.5$$

(*b*) Expected number of customers in the bank are given by

$$\overline{L_S} = \frac{\lambda}{\mu - \lambda} = \frac{6}{6} = 1$$

(*c*) Expected time to be spent in the bank is given by

$$\overline{W_S} = \frac{1}{\mu - \lambda} = \frac{1}{12 - 6} = 1/6 \text{ hour} = 10 \text{ minutes}$$

So far, we have discussed a system consisting of single queue and single service counter. A system may have a single queue and multiple service counters as shown in Fig. 7.4. So far, we assumes that arrival at tern follows Poisson's distribution and time for servicing follows an exponential distribution. But these are not the only patterns of arrival. Arrival may be totally irregular, with time as well as with day. For examples in banks, crowd is much more on Saturdays than on other days, and thus Poisson's pattern fails. Although mathematical models have been developed for different arrival patterns, yet generally industrial and business units adopt simulation techniques for this. In the next section we will discuss simulation model for single server, single queue.

7.5 SIMULATION OF A SINGLE SERVER QUEUE

In this section, we will construct a simulation model for single queue and a single server, say a machine shop. In a real life dynamic system, time flow is an essential element. Whether it is a simulation model of queuing system, manufacturing system of inventory control, many parameters in these are function of time. Thus time flow mechanism is an essential part in a simulation model. There of two basic ways of incrementing time in a simulation model as.

(*a*) **Fixed Time Increment:** In fixed time increment model, also called **Time Oriented Simulation**, events are recorded after a fixed interval of time, which is constant during the simulation period. After the end of each interval, it is noted how many customers have arrived in a queue, and how many have left the server after being served. Attempt in this system is to keep time interval as small as possible, so that minor details of model are monitored. Possible in one time interval, only one customer arrives and only one leaves. Fixed time increment simulation is generally preferred for continuous simulation. Numerical methods, where time is taken as independent variable are one such example.

(*b*) **Next Event Increment Simulation:** This method is also called **Event Oriented Simulation**. In this system, time is incremented when an event occurs. For example in queuing, when a customer arrives, clock is incremented by his arrival time. In such case time period for simulation may be stochastic.

Let us consider an example where a factory has a large number of semiautomatic machines. Out of these machines, none of the machine fails on 50% of the days, whereas on 30% of the days, one machine fails and on 20% of the days, two can fail. The maintenance staff of the factory, can repair 65% of these machines in one day depending on the type of fault, 30% in two days and 5% remaining in three days. We have to simulate the system for 30 days duration and estimate the average length of queue, average waiting time, and the server loading i.e., the fraction of time for which server is busy.

The given system is a single server queuing model. Arrival here is failure of the machines and, while maintenance is the service facility. There is no limit on the number of machines, that is queue length is infinite. Thus

Expected arrival rate $= 0 \times 0.5 + 1 \times 0.3 + 2 \times 0.2 = 0.7$ per day

And

Average service time is $= 1 \times 0.65 + 2 \times 0.3 + 3 \times 0.05 = 1.4$ days

Therefore

Expected service rate $= 1/1.4 = 0.714$ machines per day

The expected rate of arrival is slightly less than the expected servicing rate and hence the system can reach a steady state. For the purpose of generating the arrivals per day and the services completed per day the given discrete distribution will be used.

We generate uniform random numbers between 0 and 1 to generate the failure of machines (arrival of machines) as under.

$$
\begin{aligned}
0.0 < r \le 0.5 \text{ arrival} &= 0 \text{ machine} \\
0.5 < r \le 0.8 \text{ arrival} &= 1 \text{ machine} \\
0.8 < r \le 1.0 \text{ arrival} &= 2 \text{ machines}
\end{aligned}
\qquad \text{...(7.21)}
$$

Similarly, we generate random numbers for servicing as

$$
\begin{aligned}
0.0 < r \le 0.65 \text{ service time} &= 1 \text{ day} \\
0.65 < r \le 0.95 \text{ service time} &= 2 \text{ days} \\
0.95 < r \le 1.0 \text{ service time} &= 3 \text{ days}
\end{aligned}
\qquad \text{...(7.22)}
$$

We can simulate this system by taking one day as the unit of time. Timer is set to zero i.e., at time $t = 0$, queue is empty and server is idle having no job.

Day One: On day one, queue as well as server has no customer which is shown as line one in Table 7.1 with timer = 0. In order to see whether any machine comes for repair or not, we generate a uniform random number between 0 and 1. Let this number be $r = 0.413$, which being less than 0.5, from (7.21) we find that no new machine arrives for repair and thus server remains idle.

Day Two: Day is incremented by one, and timer shows 1. By this time server's idle time is one day and waiting time is also one day. Again a random number is generated to find new arrival. Let this number be $r = 0.923$. From equation (7.21), we find that two machines arrive. Since server is idle one goes for service and other waits in queue. For simulating service time we again generate a uniform random number and let this number be $r = 0.516$. From equation (7.22) we observe that for this random number repair time (service time) is one day.

Day Three: The situation on third day is: Idle time = 1day, queue = 1, and waiting time = 1 day. We increment day by one and timer becomes 3. Random number of arrival is .571 and thus one customer comes, and one in queue goes to server. Random number for server is .718 and thus service time for this machine is 2 days. Today's customer is in queue. Waiting time becomes 2 days for this customer who is in queue as server is serving second customer of day 2. Thus waiting time becomes 2. The process goes on like this.

It is to be noted that when service time is one day, means each machine takes one day for repair and when service time is 2 days, each machine takes 2 days to be served.

Results of the simulation are given in the Table 7.1.

Table 7.1: Simulation results

Timer	Random number	Arrival	Queue	Random number	Service time	Idle time	Waiting time
0		0	0		0	0	0
1	.413	0	0		0	0	0
2	.923	2	1	.516	1	1	1
3	.517	1	1	.718	2	1	2
4	.430	0	1		1	1	3
5	.394	0	0	.311	1	1	3
6	.165	0	0		0	2	3
7	.531	1	0	.955	3	2	3
8	.901	2	2	.	2	2	5
9	.722	1	3		1	2	8
10	.155	0	2	.321	1	2	10
11	.700	1	2	.711	2	2	12
12	.158	1	3		1	2	15
13	.721	0	2	.110	1	2	17
14	.871	2	3	.461	1	2	20
15	.677	1	3	.463	1	3	23
16	.469	0	2	.631	1	2	25

Contd...

17	.791	1	2	.145	1	2	27
18	.261	0	1	.801	2	2	28
19	.112	0	1		1	2	29
20	.061	0	0	.081	1	2	29
21	.461	0	0		0	3	29
22	.131	0	0		0	4	29
23	.912	1	1	.161	1	4	30
24	.456	1	1	.881	2	4	31
25	.761	2	2		1	4	33
26	.123	1	1	.531	1	4	34
27	.484	0	0	.981	3	4	34
28	.7101	1	1		4	4	35
29	.533	2	2		1	4	37
30	.901	3	3	.811	2	4	40

7.5.1 An Algorithm for Single Queue-single Server Model

An algorithm of computer program for **single queue-single server** simulation model is given below (Narsingh Deo [40]).

We simulate the arrival and servicing of N customers by a single server. Let these customers be marked 1, 2, 3,..., N. Let AT_k denotes the time gap between the arrivals of the $-(k-1)$ customer and k-th customer in the system. These times will be generated, as samples from some specified probability distribution (say exponential), by means of an appropriate random number generator. Similarly ST_k be the service time of the k-th customer, where $k = 1, 2, 3..., N$. The service time are also generated by random number of some specified probability distribution function. Let CAT_K be the cumulative arrival time of the k-th customer.

It is assumed that initially there is no customer in the queue and first customer directly goes to machine at time $t = 0$. After machining, this part goes out of system at time ST_1. Now second customer will arrive at

$$CAT_2 = AT_2$$

If the service time of first customer is greater than the arrival time of second customer ($ST_1 > CAT_2$), then second customer has to wait for time say $WT_2 = ST_1 - CAT_2$, *where* WT_K *is the wait time in the queue for* k-th *customer*. Thus a queue of length one *customer* will be formed. But if ($ST_1 < CAT_2$), then the server will be idle by the time *customer* number two arrives and queue will have no item. Thus idle waiting time for server for second item will be,

$$IDT_2 = CAT_2 - ST_1$$

where IDT_K denotes the idle time of the server awaiting for k-th item.

Let at time t, $(i - 1)$ customers have arrived into the machine shop and $(j - 1)$ customer have departed, and i-th customer is about to come and j-th customer is due to depart. If there are some customers in the queue then

$$1 \leq j \leq i \leq N$$

Thus the queue length is $(i - j - 1)$, if $i > j$. Now next arrival time NAT = CAT_i and next departure time (NDT), i.e., the cumulative departure time CDT_j of the j-th item is given by,

NDT = CDT_j = cumulative arrival time of j + waiting time of j + machining time of j.

$$= CAT_j + WT_j + ST_j.$$

We must now determine which event would take place-whether i would arrive first or j would depart first. This is decided by comparing NAT with NDT. Now if NAT < NDT, an arrival will take prior to departure and queue length will increase by one. If NAT > NDT, and $(i - j - 1)$ is also positive than a departure will take first and queue length will decrease by one. In both these cases there is no idle time for the server. However if NAT > NDT and the queue length is zero, then the server will be idle waiting for the i-th item for the duration IDT_i = (NAT – NDT). Third case will be when NAT = NDT, implies then the next arrival and departure take place simultaneously and there is no change in the queue length.

This next event simulation procedure for this simple queuing situation is shown in the flow chart (Fig. 7.3). In the flow chart, the inter arrival times AT_i's and the service times ST_i's can be generated by calling the suitable subroutines. From the inter arrival times, the cumulative arrival times are easily calculated using the relation,

$$CAT_k = CAT_{k-1} + AT_k,$$

and $$CAT_1 = AT_1 = 0$$

The event times are indicated by the variable time CLOCK.

With this algorithm, one can compute various statistics. This system is demonstrated by following example.

Example 7.5: The time between mechanic request for tools in a large plant is normally distributed with mean of 10 minutes and standard deviation of 1.5 minutes. Time to fill requests is also normal with mean of 9.5 minutes and standard deviation of 1.0 minute. Simulate the system for first 15 requests assuming a single server. Determine the average waiting time of the customers and the percentage capacity utilization. Develop a computer simulation program and obtain the said results for 1000 arrivals of customers.

Solution: The normally distributed times, the inter arrival times as well as the service times can be generated by using normal random number tables (see Appendix-9.1). Normal variable with given mean (μ) and standard deviation (σ) is given by,

$$x = \mu + \sigma \cdot (r')$$

where r' is normal random number with mean = 0 and standard deviation unity.

Simulation of the given case is carried out in the Table 7.2. First column gives arrival number. Second column gives the normal random number generated by Mean Value Theorem, for generating Inter Arrival Times (IAT). Third column gives inter arrival time, fourth gives cumulative arrival times, fifth is time for beginning the service, sixth column again is normal random number for service counter, seventh column is service time, eighth column is service end time, ninth column is customer waiting time and tenth column is Server Idle time.

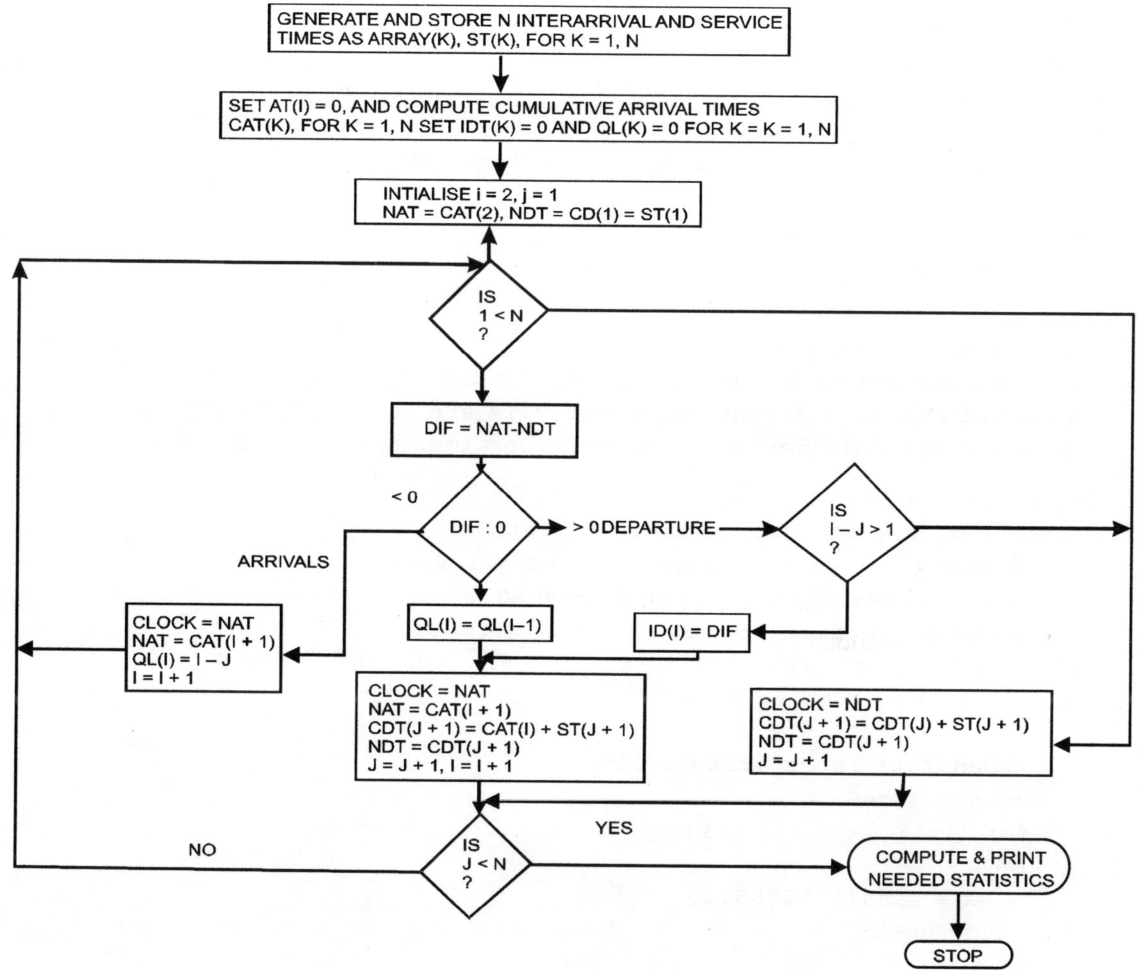

Fig. 7.3: *Flow chart of single queue-single server.*

For any arrival i, the service begins (SB) either on its arrival cumulative time or the service ending time of previous arrival, which ever is latest.

$$\text{SB}(i) = \text{Max}[\text{SE}(i-1), \text{CAT}(i)]$$

Service ending time is,

$$\text{SE}(i) = \text{SB}(i) + \text{ST}(i)$$

Waiting time for customer is,

$$\text{WT}(i) = \text{SE}(i-1) - \text{CAT}(i), \text{ if possible}$$

Service idle time is,

$$\text{IT}(i) = \text{CAT}(i) - \text{SE}(i-1), \text{ if possible}$$

Program for this simulation problem is given below. At the end of 10 simulations,

Average waiting time = 0.2669

Percentage capacity utilization = 80.88

Program 7.1: Single Queue Simulation

```
// Single queue simulation
#include <iostream.h>
#include <fstream.h>
#include <stdlib.h>//contains rand() function
#include <math.h>
#include <conio.h>
#include<iomanip.h>

void main(void)
{ /* Single server queue:
  arrival and service times are normally distributed.
  mean and standard deviation of arrivals are 10  and 1.5 minutes.
  mean and standard deviation of service times are 9.5 and 1.0 */

int i,j,run = 10;
double x,x1,x2, st, awt, pcu, wt=0, iat=0,it;
double mean=10., sd=1.5, mue=9.5, sigma=1.0;
double sb=0.,se=0.,cit=0.,cat=0.,cwt=0.;
    ofstream outfile("output.txt",ios::out);
outfile<<"\n i  r'  IAT   CAT   SB   r'  ST   SE   WT    IT\n";
for (j = 1; j <= run; ++ j)
{
    //Generate inter arrival time
    double sum=0;
    for (i=1; i < = 12; ++i)
    {
        x = rand()/32768.0;
     sum=sum+x;
    }
    x1=mean+sd*(sum-6.);
    iat= x1;
    //cout<<"iat="<<iat;
    cat=cat+iat;
    //cout<<"cat="<<cat;
    if(cat<=se)
    {
        sb=se;
        wt=se-cat;
        cwt=cwt+wt;
    //  cout<<"cwt="<<cwt;
        }
    else
```

```
        {
            sb=cat;
            it=sb-se;
            cit=cit+it;
        }

    //generate service time
    sum=0.;
        for(i=1; i<=12;++i)
        {x=rand()/32768.;
            sum=sum+x;
            x2=mue+sigma*(sum-6.);
            st=x2;
            se=sb+st;
        }
    outfile<<j<<'\t'<<setprecision(4)<<x1<<'\t'<<setprecision(4)<<iat<<'\t'<<setprecision
(4)<<cat<<'\t'<< setprecision (4)<<sb<<'\t'<< setprecision (4)<<x2<<'\t'<<
setprecision (4)<<st<<'\t'<< setprecision(4)<<se<<'\t'<<setprecision(4)
<<wt<<'\t'<<setprecision (4)<<it<<"\n";
    }
        awt=cwt/run;
        pcu=(cat-cit)*100./cat;
        outfile<<"Average waiting time\n";
        outfile<<awt;
        outfile<<"Percentage capacity utilization\n"<<pcu;
    }
```

Output of this program is given below. Ten columns are respectively number of arrival, *I*, random number *r′*, inter arrival time IAT, cumulative arrival time CAT, time at which service begins SB, random number for service time *r′*, service time ST, time for service ending SE, waiting time WT, and idle time IT.

Table 7.2: Output of program

1	2	3	4	5	6	7	8	9	10
i	*r′*	*IAT*	*CAT*	*SB*	*r′*	*ST*	*SE*	*WT*	*IT*
1	10.72	10.72	10.72	10.72	7.37	7.37	18.09	0	10.72
2	8.493	8.493	19.21	19.21	10.77	10.77	29.98	0	1.123
3	11.68	11.68	30.9	30.9	9.021	9.021	39.92	0	0.913
4	10.74	10.74	41.63	41.63	9.469	9.469	51.1	0	1.714
5	9.095	9.095	50.73	51.1	9.905	9.905	61.01	0.374	1.714
6	9.972	9.972	60.7	61.01	9.988	9.988	70.99	0.306	1.714
7	11.35	11.35	72.05	72.05	9.161	9.161	81.21	0.306	1.057

Contd...

8	10.74	10.74	82.79	82.79	7.081	7.081	89.87	0.306	1.58
9	9.19	9.19	91.98	91.98	10.53	10.53	102.5	0.306	2.11
10	8.542	8.542	100.5	102.5	10.85	10.85	113.4	1.989	2.11

The problem can be extended to multiple servers with single queue and is given in next section.

Example 7.6: Simulate an $M/M/1/\infty$ queuing system with mean arrival rate as 10 per hour and the mean service rate as 15 per hour, for a simulation run of 3 hours. Determine the average customer waiting time, percentage idle time of the server, maximum length of the queue and average length of queue.

Solution: The notation $M/M/1/\infty$ is Kendal's notation of queuing system which means, arrival and departure is exponential, single server and the capacity of the system is infinite.

Mean arrival rate is = 10 per hrs or 1/6 per minute.

Mean departure rate is = 15 per hrs or 1/4 per minute.

In actual simulation the inter arrival times and service times are generated, using exponential random numbers. Exponential random number R is given by,

$$R = -(1/\tau) \ln (1 - r)$$

where τ is arrival rate. Below we give a program for generating these numbers.

Program 7.2: Single Queue - Single Server Simulation (Example 7.6)

```
// Single queue single servers simulation

#include <iostream.h>
#include <fstream.h>
#include <stdlib.h>//contains rand() function
#include <math.h>
#include <conio.h>
#include<iomanip.h>

void main(void)
{
/* Single  server single queue:
  arrival and service times are exponentially distributed.
   iat is inter arrival time.
  count a and count b are counters for server A and B.
  qa(qb) are que length of component A(B).
*/
     //M/M/1/infinity queue
     double r,iat;
double mue=1/6., lemda=1/5., run=180.;
double
```

```
clock=0.00,se=0.00,sb=0.00,nat=0.00,cit=0.00,cwt=0.00,st=0.00,it=0.00,wt=0.00;
int q=0,cq=0,k,count=0,qmax=100;

ofstream outfile("output.txt",ios::out);

    outfile<<"\n\n CLOCK IAT NAT QUE SB ST SE IT WT CIT CWT\n";

        r=rand()/32768.;
        cout<<"rand="<<r<<endl;
        iat=(-1./mue)*log(1-r);
        nat = nat+iat;
        count=count+1;
    // cout<<nat<<count<<endl;
    // getch();
    while(clock<=run) {
        if(q>qmax) qmax=q;
outfile<<setprecision(4)<<clock<<'\t'<<setprecision(4)<<iat<<'\t'<<
setprecision(4)<<nat<'\t'<<setprecision(4)<<q<<'\t'<<setprecision(4)<<sb<<'
\t'<<setprecision(4)<<st<<'\t'<<setprecision(4)<<se<<'\t'<<setprecision(4)<<it<<'\
t'<<setprecision(4)<<wt<<'\t'<<setprecision(4)<<cit<<'\t'<<setprecision(4)<<cwt<<endl;

        if(nat>=se){
           if(q>0){
              wt=q*(se-clock);
              cwt=cwt+wt;
              q=q-1;
              clock=se;
           }
           else
           { clock=nat;
              r=rand()/32768.;
          iat=(-1/mue)*log(1-r);
        nat = nat=iat;
        count=count+1;
           }
           sb=clock;
           it=clock- se;
           cit=cit+it;

         r=rand()/32768.;
          st=(-1/lemda)*log(1-r);
          se=sb+st;
        }
        else
```

```
        {
            wt=q*iat;
            cwt=cwt+wt;
            clock=nat;
            q=q + 1;
        r=rand()/32768.;
           st=(-1/lemda)*log(1-r);
           nat=nat+iat;
           count=count+1;
    }

  }
 outfile<<"Elapsed time="<<clock<<"Number of arrivals="<<count<<endl;
 outfile<<"Average waiting time/arrival="<<cwt/count<<endl;
 outfile<<"Average server idle time/arrival="<<cit*100./clock<<endl;
 outfile<<"Qmax="<<qmax<<endl;
 cout<<"\n any 'digit";
 cin >> k;
 }
```

Table 7.3: Output of the program

CLOCK	IAT	NAT	QUE	SB	ST	SE	IT	WT	CIT	CWT
0.00	0.0075	0.007	0	0.00	0.00	0.00	0.00	0.00	0.00	0.00
0.007	4.975	4.975	0	0.0075	1.074	1.082	0.0075	0.00	0.0075	0.00
4.975	9.924	9.924	0	4.975	4.397	9.372	3.893	0.00	3.901	0.00
9.924	3.922	3.922	0	9.924	2.156	12.08	0.552	0.00	4.453	0.00
3.922	3.922	7.844	1	9.924	11.31	12.08	0.552	0.00	4.453	0.00
7.844	3.922	11.77	2	9.924	8.653	12.08	0.552	3.922	4.453	3.922
11.77	3.922	15.69	3	9.924	6.864	12.08	0.552	7.844	4.453	11.77
12.08	3.922	15.69	2	12.08	0.9564	13.04	0.00	0.9428	4.453	12.71
13.04	3.922	15.69	1	13.04	9.792	22.83	0.00	1.913	4.453	14.62
15.69	3.922	19.61	2	13.04	6.198	22.83	0.00	3.922	4.453	18.54
19.61	3.922	23.53	3	13.04	3.603	22.83	0.00	7.844	4.453	26.39
22.83	3.922	23.53	2	22.83	1.812	24.64	0.00	9.657	4.453	36.04
23.53	3.922	27.45	3	22.83	0.075	24.64	0.00	7.844	4.453	43.89
24.64	3.922	27.45	2	24.64	0.473	25.12	0.00	3.327	4.453	47.21
25.12	3.922	27.45	1	25.12	2.266	27.39	0.00	0.9585	4.453	48.17
27.39	3.922	27.45	0	27.39	0.797	28.18	0.00	2.266	4.453	50.44
27.45	3.922	31.38	1	27.39	0.907	28.18	0.00	0.00	4.453	50.44
28.18	3.922	31.38	0	28.18	22.32	50.51	0.00	0.7293	4.453	51.17

Contd...

31.38	3.922	35.3	1	28.18	2.95	50.51	0.00	0.00	4.453	51.17
35.3	3.922	39.22	2	28.18	0.634	50.51	0.00	3.922	4.453	55.09
39.22	3.922	43.14	3	28.18	0.023	50.51	0.00	11.77	4.453	74.7
47.06	3.922	50.99	5	28.18	2.373	50.51	0.00	15.69	4.453	90.39
50.51	3.922	50.99	4	50.51	3.793	54.3	0.00	17.22	4.453	107.6
50.99	3.922	54.91	5	50.51	4.233	54.3	0.00	15.69	4.453	123.3
54.3	3.922	54.91	4	54.3	4.603	58.9	0.00	16.58	4.453	139.9
54.91	3.922	58.83	5	54.3	4.672	58.9	0.00	15.69	4.453	155.6
58.83	3.922	62.75	6	54.3	0.909	58.9	0.00	19.61	4.453	175.2
58.9	3.922	62.75	5	58.9	5.439	64.34	0.00	0.4479	4.453	175.6
62.75	3.922	66.67	6	58.9	2.996	64.34	0.00	19.61	4.453	195.2
64.34	3.922	66.67	5	64.34	2.17	66.51	0.00	9.549	4.453	204.8
66.51	3.922	66.67	4	66.51	0.294	66.81	0.00	10.85	4.453	215.6
66.67	3.922	70.59	5	66.51	4.678	66.81	0.00	15.69	4.453	231.3
66.81	3.922	70.59	4	66.81	7.646	74.45	0.00	0.6668	4.453	232
70.59	3.922	74.52	5	66.81	8.112	74.45	0.00	15.69	4.453	247.7
74.45	3.922	74.52	4	74.45	3.668	78.12	0.00	19.29	4.453	267
74.52	3.922	78.44	5	74.45	1.797	78.12	0.00	15.69	4.453	282.6
78.12	3.922	78.44	4	78.12	10.44	88.56	0.00	18.02	4.453	300.7
78.44	3.922	82.36	5	78.12	6.485	88.56	0.00	15.69	4.453	316.4
82.36	3.922	86.28	6	78.12	15.6	88.56	0.00	19.61	4.453	336
86.28	3.922	90.2	7	78.12	13.00	88.56	0.00	23.53	4.453	359.5
88.56	3.922	90.2	6	88.56	3.875	92.43	0.00	15.91	4.453	375.4
90.2	3.922	94.13	7	88.56	0.768	92.43	0.00	23.53	4.453	398.9
92.43	3.922	94.13	6	92.43	3.10	95.53	0.00	15.59	4.453	414.5
94.13	3.922	98.05	7	92.43	1.341	95.53	0.00	23.53	4.453	438.1
95.53	3.922	98.05	6	95.53	9.91	105.4	0.00	9.836	4.453	447.9
98.05	3.922	102.00	7	95.53	1.176	105.4	0.00	23.53	4.453	471.4
102.00	3.922	105.9	8	95.53	7.562	105.4	0.00	27.45	4.453	498.9
105.4	3.922	105.9	7	105.4	9.278	114.7	0.00	27.77	4.453	526.7
105.9	3.922	109.8	8	105.4	28.67	114.7	0.00	27.45	4.453	554.1
109.8	3.922	113.7	9	105.4	40.0	114.7	0.00	31.38	4.453	585.5
113.7	3.922	117.7	10	105.4	4.727	114.7	0.00	35.3	4.453	620.8
114.7	3.922	117.7	9	114.7	2.491	117.2	0.00	9.834	4.453	630.6
117.2	3.922	117.7	8	117.2	1.548	118.8	0.00	22.42	4.453	653
117.7	3.922	121.6	9	117.2	1.764	118.8	0.00	31.38	4.453	684.4
118.8	3.922	121.6	8	118.8	9.167	127.9	0.00	9.904	4.453	694.3
121.6	3.922	125.5	9	118.8	0.1201	127.9	0.00	31.38	4.453	725.
125.5	3.922	129.4	10	118.8	2.357	127.9	0.00	35.3	4.453	761
127.9	3.922	129.4	9	127.9	0.486	128.4	0.00	24.23	4.453	785.2
128.4	3.922	129.4	8	128.4	5.653	134.1	0.00	4.374	4.453	789.6

Contd...

129.4	3.922	133.3	9	128.4	0.2893	134.1	0.00	31.38	4.453	821
133.3	3.922	137.3	10	128.4	0.0441	134.1	0.00	35.3	4.453	856.3
134.1	3.922	137.3	9	134.1	12.55	146.6	0.00	7.187	4.453	863.5
137.3	3.922	141.2	10	134.1	1.614	146.6	0.00	35.3	4.453	898.8
141.2	3.922	145.1	11	134.1	1.593	146.6	0.00	39.22	4.453	938
145.1	3.922	149.0	12	134.1	4.432	146.6	0.00	43.14	4.453	981.1
146.6	3.922	149.0	11	146.6	5.875	152.5	0.00	18.06	4.453	999.2
149.0	3.922	153.0	12	146.6	9.088	152.5	0.00	43.14	4.453	1042
152.5	3.922	153.0	11	152.5	6.482	159	0	41.49	4.453	1084
153.0	3.922	156.9	12	152.5	3.317	159	0	43.14	4.453	1127
156.9	3.922	160.8	13	152.5	1.149	159	0	47.06	4.453	1174
159.0	3.922	160.8	12	159.0	6.807	165.8	0.00	27.24	4.453	1201
160.8	3.922	164.7	13	159.0	3.16	165.8	0.00	47.06	4.453	1248
164.7	3.922	168.6	14	159.0	3.062	165.8	0.00	50.99	4.453	1299
165.8	3.922	168.6	13	165.8	14.89	180.7	0.00	14.82	4.453	1314
168.6	3.922	172.6	14	165.8	6.821	180.7	0.00	50.99	4.453	1365
172.6	3.922	176.5	15	165.8	0.573	180.7	0.00	54.91	4.453	1420
176.5	3.922	180.4	16	165.8	4.569	180.7	0.00	58.83	4.453	1479

Elapsed time = 180.4, Number of arrivals = 50
Average waiting time/arrival = 30.83
Average server idle time/arrival = 2.468
Q_{max} = 16

7.5.2 Infinite Queue-infinite Source, Multiple-server Model

There may be cases when queue is single but servers are more than one (Fig. 7.3). Let us generalize the concepts developed in earlier sections to single queue two server problem. We assume here that,

(*a*) s denotes number of servers in the system,
(*b*) Each server provides service at the same constant rate μ,
(*c*) Average arrival rate for all n customers is constant λ,
(*d*) $\lambda < s\mu$.

When the mean service rate per busy server is μ, the overall mean service rate for n busy servers will be $n\mu$. With these assumptions probability that there are $(n + 1)$ customers in the system at time t is given by putting $\mu = (n + 1)\mu$ in equation (7.9), one gets

$$P_{n+1}(t) + \frac{\lambda}{(n+1)\mu} P_{n-1}(t) - \left(\frac{\lambda}{(n+1)\mu} + 1\right) \cdot P_n(t) = 0$$

This equation can be written as,

$$(n+1)\mu P_{n+1}(t) = [(\lambda + \mu)P_n(t) + (n\mu \cdot P_n(t) - \lambda P_{n-1}(t))]$$

From this equation, we can get by putting $n = 1, 2, 3, \ldots, s$

$$P_1 = \left(\frac{\lambda}{\mu}\right) P_0$$

$$P_2 = \left(\frac{\lambda}{2\mu}\right) P_1 + \frac{1}{2\mu}(\mu P_1 - \lambda P_0) = \frac{1}{2}\left(\frac{\lambda}{\mu}\right)^2 P_0$$

$$P_3 = \left(\frac{\lambda}{3\mu}\right) P_2 + \frac{1}{3\mu}(2\mu P_2 - \lambda P_1) = \frac{1}{3!}\left(\frac{\lambda}{\mu}\right)^3 P_0$$

Thus generalizing above results, the probability that there are n customers in the system, For $n \geq s$ is given by,

$$P_n = \frac{1}{n!}\left(\frac{\lambda}{\mu}\right)^n P_0, \text{ for } n = 0, 1, 2, \ldots, s \qquad \ldots(7.23)$$

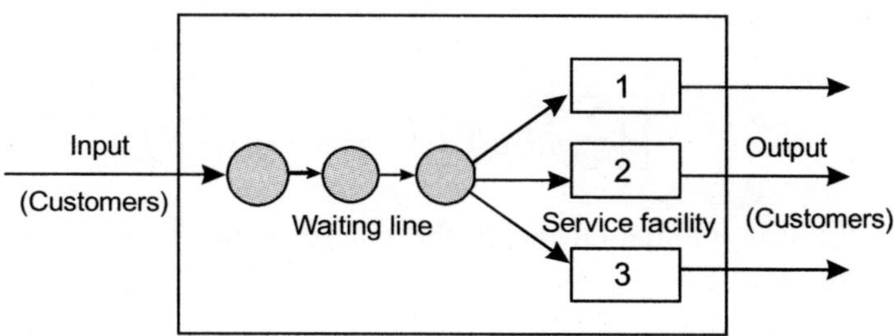

Fig. 7.4: *Single queue multiple counters.*

When there are s servers and $n \geq s$, then

$$P_n = \frac{1}{s!}\left(\frac{\lambda}{\mu}\right)^s \left(\frac{\lambda}{s\mu}\right)^{n-s} P_0, \text{ for } n > s \qquad \ldots(7.24)$$

$$\therefore \qquad P_n = \frac{1}{s!s^{n-s}}\left(\frac{\lambda}{\mu}\right)^n P_0, \text{ for } n > s$$

This means that there are s customers in servers and $(n - s)$ in the queue waiting for the service. There arrival rate thus is λ and departure rate is $s\mu$.

Now probability that there are at least s customers in the system is,

$$P(n \geq s) = \sum_{n=s}^{\infty} P_n$$

$$= \sum_{n=s}^{\infty} \frac{1}{s!s^{n-s}}\left(\frac{\lambda}{\mu}\right)^n P_0$$

$$= \frac{1}{s!}\left(\frac{\lambda}{\mu}\right)^s \sum_{n=s}^{\infty}\left(\frac{\lambda}{s\mu}\right) P_0$$

$$= \frac{1}{s!}\left(\frac{\lambda}{\mu}\right)^s \sum_{n=s}^{\infty}\left(\frac{\lambda}{s\mu}\right) P_0 \qquad \text{...(7.24}a\text{)}$$

Since sum of all the probabilities from $n = 0$ to $n = \infty$ is equal to unity, therefore,

$$P_0 = \frac{1}{\left[1 + \left\{\sum_{n=1}^{n=s} \frac{1}{n!}\left(\frac{\lambda}{\mu}\right)^n\right\} + \frac{\left(\frac{\lambda}{\mu}\right)^s}{s!} \sum_{n=s+1}^{\infty} \left(\frac{\lambda}{s\mu}\right)^{n-s}\right]} \qquad \text{...(7.25)}$$

$$P_0 = \frac{1}{\left[\left\{\sum_{n=0}^{s} \frac{1}{n!}\left(\frac{\lambda}{\mu}\right)^n\right\} + \frac{1}{s!(1-\lambda/\mu s)}\left(\frac{\lambda}{\mu}\right)^s\right]} \qquad \text{...(7.26)}$$

where $n = 0$, term in the last summation yields the correct value of 1 because of the convention that $n! = 1$ when $n = 0$.

Average length of the queue (L_Q) is,

$$L_Q = \sum_{n=s}^{\infty} (n-s) P_n$$

$$= \sum_{j=0}^{\infty} j P_{s+j}$$

$$= \sum_{j=0}^{\infty} j \frac{(\lambda/\mu)^s}{s!}\left(\frac{\lambda}{s\mu}\right)^j P_0$$

$$= P_0 \frac{(\lambda/\mu)^s}{s!} \rho \sum_{j=0}^{\infty} \frac{d}{d\rho}(\rho^j) \qquad \text{...(7.27)}$$

which is same as

$$\frac{\left(\frac{\lambda}{\mu}\right)^s \rho P_0}{s!(1-\lambda/\mu s)^2}$$

Average number of customers in the system are equal to average number of customers in the queue plus average number of customers in the server i.e.,

$$\overline{L_S} = \overline{L_Q} + \frac{\lambda}{\mu} \quad ...(7.28)$$

$$\overline{W_S} = L_S/\lambda \text{ and } \overline{W_Q} = \overline{L_Q}/\lambda \quad ...(7.29)$$

where L_Q is the average number of customers waiting in the queue to be served.

Similarly probability that waiting time is greater than t is,

$$P(T > t) = e^{-\mu t}\left\{1 + \frac{(\lambda/\mu)^s P_0[1 - e^{-\mu t(s-1-\lambda/\mu)}]}{s!(1-\lambda/\mu s)(s-1-\lambda/\mu)}\right\} \quad ...(7.30)$$

Example 7.7: In an office there are two typist in a typing pool. Each typist can type an average of 6 letters/hour. If letters arrive for typing at the rate of 10 letters/hour, calculate,

(*a*) What fraction of the time are all the two typists busy?

(*b*) What is the average number of letters waiting to be typed?

(*c*) What is the waiting time for a letter to be typed?

Solution:

Here λ = 10 letters/hour

μ = 6 letters/hour

$s = 2$

(*a*) Here we have to find the probability that there are at least two customers in the system, that is, we want that $P(n \geq 2)$. To get this

$$P_0 = \frac{1}{\left[1 + (\lambda/\mu) + \frac{1}{2}(\lambda/\mu)^2 (1/(1-\lambda/2\mu))\right]}$$

$$= \frac{1}{\left[1 + (10/6) + \frac{1}{2}(10/6)^2 (1/(1-10/12))\right]}$$

$$= 0.06688$$

Therefore

$$P(n \geq 2) = \frac{(\lambda/\mu)^s P_0}{s!(1-\lambda/\mu s)}$$

$$= \frac{(1.67)^2 \times 0.06688}{2(1-10/12)}$$

$$= \frac{2.7889 \times 0.06688}{0.234} = 0.7971$$

this means 79.71% of the time both the typists are busy.

Note that the probability of one letter in the system (none waiting to be typed and one being typed) is

$$P_1 = \frac{2.7889 \times 0.06688}{0.234} = 0.7971$$

(*b*) Average number of letters waiting to be typed are given by,

$$L_q = \frac{(\lambda/\mu)^{s+1} P_0}{s \cdot s!(1-\lambda/\mu s)^2}$$

$$= \frac{(10/6)^3\, 0.6688}{2 \cdot 2!(1-10/12)^2}$$

$$= 1.15999$$

(*c*) Average time a letter spends in the system (waiting and being typed) is given as,

First we calculate average number of letters in the system i.e.,

$$L = L_q + \frac{\lambda}{\mu} = 1.15999 + 1.6667 = 2.8267$$

Therefore,

$$W = L/\lambda = 2.8267/10 = 0.283 \text{ hr} = 13 \text{ minutes}$$

7.5.3 Simulation of Single Queue Multiple Servers

Simulation of multiple server queue is very important seeing its application in day to day life. Let us consider a case of bank where there are two service counters. Customers arrive in the bank according to some probability distribution for arrival times. When a customer enters the bank, he checks whether a counter is free or not. If a counter is free, he will go to that counter, else he will stand in queue of one of the counter, preferably in a smaller queue increasing the queue length by one. Customer is attended at the service counter as per first come first served rule. The service time from each service counter can be viewed as independent sample from some specified distribution. It is not necessary that inter arrival time or service time be exponential. Queuing system although simple to describe, is difficult to study analytically. In such cases simulation is the only alternative.

Let us first simulate this situation in tabular form. We assume that six customers come to bank at times 0, 9, 13, 22, 26, 33 minutes. The simulation starts at time zero. At this time there are no customers in the system. Column 1 is simply the serial number of the customer. Second column is inter arrival time of *k*-th customer and is denoted by AT_k. That is AT_k is the inter arrival time between the arrival of *k*-th and (*k* – 1)-th customer. This means,

$$AT_1 = 0$$

Third column is cumulative arrival time CAT_k of *k*-th customer i.e.,

$$CAT_k = CAT_{k-1} + AT_k$$

and

$$CAT_1 = AT_1 = 0$$

Table 7.4: Simulation results

			Counter 1			Counter 2				
k	AT_k	CAT_k	$ST_{k,1}$	$CDT_{k,1}$	$IDT_{k,1}$	$ST_{k,2}$	$CDT_{k,2}$	$IDT_{k,2}$	WT_k	QL_k
1	0	0	10	10	0	–	–	–	0	0
2	9	9	–	–	–	7	16	9	0	0
3	4	13	12	25	3	–	–	–	0	0
4	9	22	–	–	–	20	42	6	0	0
5	4	26	15	41	1	–	–	–	1	1
6	7	33	–	–	–	15	48	0	8	1

As the first customer arrives at bank, he directly goes to counter 1. Let service time of first customer at the service counter is say 10 minutes. Thus within ten minutes first customer leaves the first counter. Column four gives $ST_{k,1}$ the service time for k-th customer at column 1. Similarly column seven gives service time at counter 2. Column five gives the cumulative departure time of k-th customer from counter 1, denoted by $CDT_{k,1}$. In the sixth column is the idle time $I_DT_{k,1}$ of counter 1, waiting for k-th customer to arrive. Similarly next three columns are defined for service counter 2. The tenth column gives the waiting time (WT_k) for k-th customer in the queue. Last column gives the queue length (QL_k) immediately after the arrival of k-th customer.

At time 9 minutes second customer arrives and directly goes to second counter as first counter is busy. Let service time of second customer at the second counter is 7 minutes, it leaves the counter at 16 minutes i.e., $CDT_{2,2} = 16$. Tenth column gives idle time for second counter i.e., $IDT_{2,2} = 9$, as the second counter was idle, till the second customer arrived.

Third customer arrives at 13 minute and checks if some counter is free, otherwise he waits in the queue for counter to be free. This is done by comparing the latest values of the cumulative departure times $CDT_{1,1}$ and $CDT_{2,2}$. The smaller of the two indicates when the next facility will be free. Thus the next departure time is,

$$MNDT = \min(10,16)$$

Thus Counter one became free at time 10 and is waiting for the customer, who goes to it. Let its service time be 12 minutes, therefore departure of this customer from counter 1 will be 25 minutes. This counter remained idle for 3 minutes and queue length is zero.

Next customer arrives at time 22 minutes i.e.,

$$CAT_4 = 22. \text{ Since}$$

$$CAT_4 > MNDT(CDT_{3,1}, CDT_{2,2})$$

$$CAT_4 - MNDT = 22 - 16 = 6 = IDT_{4,2}$$

Cumulative departure time at counter number 2 of customer number 4 is given by,

$$CDT_{4,2} = CAT_4 + ST_{4,2}$$

$$= 22 + 20 = 42$$

Fifth customer arrives at 26 minutes and cumulative departure time of counter 1 is 25 minutes. This customer goes to counter one and its idle time is one minute.

Similarly when sixth customer arrives at 33 minute no counter is free. Minimum cumulative departure time is 41 of counter 1. This customer has to wait for eight minutes and then go to counter 1.

Example 7.8: In an *M*/*D*/2/3 system, the mean arrival time is 3 minutes and servers I and II take exactly 5 and 7 minutes respectively (lambda 1 and lambda 2 in the program) to serve a customer. Simulate the system for the first one hour of operation, determine the idle time of servers and waiting time of customers.

Solution: In an *M*/*D*/2/3 system, arrivals are distributed exponentially, while the service times are deterministic. Digit 2 here means there are two servers and 3 means, total number of customers in the system can not exceed three. The next customer in this case will be returned without service. The queue discipline will be taken as FIFO.

The mean inter arrival time = 3 minutes

Which means arrival rate λ = 1/3 minutes

For arrival schedule of customers, the random number x, for exponential distribution is generated as,

$$x = (-1/\lambda)\cdot\ln(1 - r)$$

where r is a uniform random number between 0 and 1.

Simulation program is given in the Computer program. It is assumed that at time zero, there is no customer in the queue. To compute the time of first arrival, we compute an exponential random number.

Program 7.3: Single Queue Two Servers Simulation

```
// Single queue two servers simulation
#include <iostream.h>
#include <fstream.h>
#include <stdlib.h>//contains rand() function
#include <math.h>
#include <conio.h>
#include<iomanip.h>
void main(void)
{ /* M/D/2/3 queuing system.*/
int k,q=0,qmax=3,count=0,counter;
double r, iat,clock=0., nat=0., wt2=0., wt1=0.,it1=0.,it2=0.,cit1=0.,
      cit2=0.;
double mean=3.,lambda1=5.,lambda2=4.,se1=0.,se2=0.,run=150;
      ofstream outfile("output.txt",ios::out);
      outfile<<"\n CLOCK IAT  NAT  SE1  SE2  QUE  COUNT
CIT1CIT2 \n";
      // Generate first arrival
         while (clock<=run){
      //Check the state of arriv7al and update que
        r = rand()/32768.0;
        iat=(-mean)*log(1-r);
         nat=nat+iat;
         se1=lambda1;//Service time taken by first server
         counter=1;//First customer has come counter=1
```

```
      outfile.precision(4);
outfile<<clock<<'\t'<<iat<<'\t'<<nat<<'\t'<<se1<<'\t'<<se2<<'\t'<<q<<'\t'
<<count<<'\t'<<cit1<<cit2<<endl;
   //it1 and it2 are idle times for two servers.
   while(clock<=run)
      {
      if(nat<=se1 && nat<=se2){
          clock=nat; q=q+1;
            r = rand()/32768.0;
           iat=(-mean)*log(1-r);
            nat=nat+iat; counter=counter+1;
      }
      else if(se1<=nat && se1<=se2) clock=se1;
      else clock=se2;

          if (q>qmax){ count=count+1;
            q=q-1;
   }

   if (q>=1 && se1<=clock) {
       it1=clock-se1;
       cit1=cit1+it1;
       se1=clock+lambda1;
       q=q-1;
   }

     if(q>=1 &&se2<=clock)
       {
          it2=clock-se2;
          cit2=cit2+it2;
          se2=clock+lambda2;
          q=q-1;
       }
      if(q==0 && se1<=clock)
      {
          clock=nat;
          it1=clock-se1;
          cit1=cit1+it1;
          se1=nat+lambda1;
          se1=nat+lambda1;
          r = rand()/32768.0;
    iat=(-mean)*log(1-r);
          nat=nat+iat;
```

```
            counter=counter+1;
        }
            if(q==0 && se2<=clock)
        {
            clock=nat;
            it2=clock-se2;
            cit2=cit2+it2;
            se2=nat+lambda2;
            r = rand()/32768.0;
      iat=(-mean)*log(1-r);
            nat=nat+iat;
            counter=counter+1;
        }
    outfile<<clock<<'\t'<<iat<<'\t'<<nat<<'\t'<<se1<<'\t'<<se2<<'
    \t'<<q<<'\t'<<count<<'\t'<<cit1<<'\t'<<cit2<<endl;

    }
    outfile.precision(4);
    outfile<<"clock="<<clock<<"cit1="<<cit1<<"cit2="<<cit2<<"counter="<<counter<<endl;
    outfile<<"Queuing system M/D/2/3"<<endl;
     outfile<<"Mean of the exponential distribution="<<mean<<endl;
     outfile<<"service time of two servers="<<lambda1<<'\t'<<lambda 2<<endl;
    outfile<<"Simulation run time="<<clock<<endl;
    outfile<<"Number of customers arrived"<<counter<<endl;
     outfile<<"Number of customers returned without service"<<count<<endl;
    outfile<<"idle time of serverI\n"<<cit1<<endl;
    outfile<<"idle time of server II\n"<<cit2<<endl;
    outfile<<"Percentage idle time of serverI\n"<<cit1*100/clock<<endl;
    outfile<<"Percentage idle time of serverII\n"<<cit2*100/clock<<endl;
            }
    cout<<"any number"<<endl;
    cin>>k;
    }
```

In the Table 7.5, first customer comes at 0.004 minute and goes directly to server1 (SE_1) whose service time is 5 minutes. In second line, second customer comes after 2.487 minutes, so net arrival time is 2.491 minutes and he goes to server 2 which is idle (cit2) for 0.004 minutes. Third customer comes at nat = 3.136. Server 1 and server 2, both are busy and hence he waits and queue length is 1. Fourth customer arrives at cumulative time 8.098 minutes, thus queue length becomes 2. But second server becomes free at 4.004 minutes and one customer goes to it at 4.004 minutes and clock time is set at 4.004 minutes and queue length at this time becomes 1. Next line shows clock time equal to 5 minutes as second customer in queue goes to first server, which becomes free at 5.00 minutes. Similarly further lines can be explained.

Table 7.5: Output of simulation

CLOCK	IAT	NAT	SE_1	SE_2	QUE	COUNT	CIT_1	CIT_2
0.00	0.004	0.004	5.00	0.00	0	0	0.00	0.00
0.004	2.487	2.491	5.00	4.00	0	0	0.00	0.004
2.491	0.644	3.136	5.00	4.004	1	0	0.00	0.004
3.136	4.962	8.098	5.00	4.004	2	0	0.00	0.004
4.004	4.962	8.098	5.00	8.004	1	0	0.00	0.004
5.00	4.962	8.098	10.00	8.004	0	0	0.00	0.004
8.098	2.638	10.74	10.00	12.1	0	0	0.00	0.09751
10.74	1.961	12.7	15.74	12.1	0	0	0.74	0.09751
12.7	1.294	13.99	15.74	16.7	0	0	0.74	0.6968
13.99	6.788	20.78	15.74	16.7	1	0	0.74	0.6968
15.74	6.788	20.78	20.74	16.7	0	0	0.74	0.6968
20.78	5.192	25.97	20.74	24.78	0	0	0.74	4.779
30.09	0.574	30.66	30.97	34.09	0	0	5.97	10.09
30.66	5.875	36.54	30.97	34.09	1	0	5.97	10.09
30.97	5.875	36.54	35.97	34.09	0	0	5.97	10.09
36.54	3.719	40.26	35.97	40.54	0	0	5.97	12.54
40.26	2.162	42.42	45.26	40.54	0	0	10.26	12.54
42.42	1.087	43.51	45.26	46.42	0	0	10.26	14.42
43.51	0.046	43.55	45.26	46.42	1	0	10.26	14.42
43.55	0.29	43.84	45.26	46.42	2	0	10.26	14.42
43.84	1.36	45.2	45.26	46.42	3	0	10.26	14.42
45.2	0.478	45.68	45.26	46.42	3	1	10.26	14.42
45.26	0.478	45.68	50.26	46.42	2	1	10.26	14.42
45.68	0.544	46.22	50.26	46.42	3	1	10.26	14.42
46.22	13.39	59.61	50.26	46.42	3	2	10.26	14.42
46.42	13.39	59.61	50.26	50.42	2	2	10.26	14.42
50.26	13.39	59.61	55.26	50.42	1	2	10.26	14.42
50.42	13.39	59.61	55.26	54.42	0	2	10.26	14.42
59.61	1.77	61.38	55.26	63.61	0	2	10.26	19.61
61.38	0.380	61.77	66.38	63.61	0	2	16.38	19.61
61.77	0.014	61.78	66.38	63.61	1	2	16.38	19.61
61.78	0.027	61.81	66.38	63.61	2	2	16.38	19.61
61.81	1.424	63.23	66.38	63.61	3	2	16.38	19.61
63.23	2.276	65.51	66.38	63.61	3	3	16.38	19.61
63.61	2.276	65.51	66.38	67.61	2	3	16.38	19.61
65.51	2.54	68.05	66.38	67.61	3	3	16.38	19.61
66.38	2.54	68.05	71.38	67.61	2	3	16.38	19.61
67.61	2.54	68.05	71.38	71.61	1	3	16.38	19.61

Contd...

68.05	2.762	70.81	71.38	71.61	2	3	16.38	19.61
70.81	2.803	73.61	71.38	71.61	3	3	16.38	19.61
71.38	2.803	73.61	76.38	71.61	2	3	16.38	19.61
71.61	2.803	73.61	76.38	75.61	1	3	16.38	19.61
73.61	0.545	74.16	76.38	75.61	2	3	16.38	19.61
74.16	3.263	77.42	76.38	75.61	3	3	16.38	19.61
75.61	3.263	77.42	76.38	79.61	2	3	16.38	19.61
76.38	3.263	77.42	81.38	79.61	1	3	16.38	19.61
77.42	1.798	79.22	81.38	79.61	2	3	16.38	19.61
79.22	1.302	80.52	81.38	79.61	3	3	16.38	19.61
79.61	1.302	80.52	81.38	83.61	2	3	16.38	19.61
80.52	0.176	80.7	81.38	83.61	3	3	16.38	19.61
80.7	2.807	83.5	81.38	83.61	3	4	16.38	19.61
81.38	2.807	83.5	86.38	83.61	2	4	16.38	19.61
83.5	4.588	88.09	86.38	83.61	3	4	16.38	19.61
83.61	4.588	88.09	86.38	87.61	2	4	16.38	19.61
86.38	4.588	88.09	91.38	87.61	1	4	16.38	19.61
87.61	4.588	88.09	91.38	91.61	0	4	16.38	19.61
88.09	4.867	92.96	91.38	91.61	1	4	16.38	19.61
91.38	4.867	92.96	96.38	91.61	0	4	16.38	19.61
92.96	2.201	95.16	96.38	96.96	0	4	16.38	20.96
95.16	1.078	96.24	96.38	96.96	1	4	16.38	20.96
96.24	6.261	102.5	96.38	96.96	2	4	16.38	20.96
96.38	6.261	102.5	101.4	96.96	1	4	16.38	20.96
96.96	6.261	102.5	101.4	101.0	0	4	16.38	20.96
102.5	3.891	106.4	101.4	106.5	0	4	16.38	22.5
106.4	9.362	115.8	111.4	106.5	0	4	21.39	22.5
115.8	7.799	123.5	111.4	119.8	0	4	21.39	31.75
125.9	0.46	126.3	128.5	129.9	0	4	33.55	37.87
126.3	1.86	128.2	128.5	129.9	1	4	33.55	37.87
128.2	0.80	129.0	128.5	129.9	2	4	33.55	37.87
128.5	0.80	129.0	133.5	129.9	1	4	33.55	37.87
129.0	5.95	134.9	133.5	129.9	2	4	33.55	37.87
129.9	5.946	134.9	133.5	133.9	1	4	33.55	37.87
133.5	5.946	134.9	138.5	133.9	0	4	33.55	37.87
134.9	0.706	135.7	138.5	138.9	0	4	33.55	38.95
135.7	4.537	140.2	138.5	138.9	1	4	33.55	38.95
138.5	4.537	140.2	143.5	138.9	0	4	33.55	38.95
140.2	5.567	145.8	143.5	144.2	0	4	33.55	40.19
163.0	24.00	187.0	150.8	167.0	0	4	35.76	58.96

State of the system at 163 minutes is,

Queuing system *M*/*D*/2/3

Mean of the exponential distribution = 3

Service time of two servers = 5 and 4

Simulation run time = 163

Number of customers arrived = 54

Number of customers returned without service = 4

Idle time of server 1 = 35.76

Idle time of server 2 = 58.96

Percentage idle time of server 1 = 21.94

Percentage idle time of server 2 = 36.18

EXERCISE

1. Is Poisson's arrival pattern for queuing is valid for all types of queues? Explain with an example.
2. How various arrival patterns are generated for the queues? Explain with examples. (PTU, 2004)
3. Simulate a queue with single queue, two servers. Make your own assumption about the arrival patterns of customers. (PTU, 2004)
4. Discuss Kendall's notation for specifying the characteristics of a queue with an example.
5. Jobs arrive at a machine shop at fixed intervals of one hour. Processing time is approximately normal and has a mean of 50 minutes per job, and a standard deviation of 5 minutes per job. Simulate the system for 10 jobs. Determine the idle time of operator and job waiting time, and waiting time of the job. Assume that the first job arrives at time zero. Use the fixed time incremental model.
6. Repeat the simulation of problem 4, by employing the next event incremental model. Use the same string of random numbers as in problem 4, and compare the results.
7. An air force station has a schedule of 15 transport flights leaving per day, each with one pilot. Three reserve pilots are available to replace any pilot who falls sick. The probability distribution for the daily number of pilots who fall sick is as follows:

Number of pilots sick	0	1	2	3	4	5
Probability	0.15	0.20	0.25	0.15	0.15	0.10

Use the Monte Carlo simulation to estimate the utilization of reserved pilots. What is the probability of canceling one flight due to non-availability of pilot? Simulate the system for 20 days, 40 days, and 60 days writing a program in C++.

8. Lallu and Ramu are the two barbers in a barber shop, they own and operate. They provide two chairs for customers who are waiting to begin a hair cut, so the number of customers in the shop varies from 0 to 4. For $n = 0, 1, 2, 3, 4$, the probability P_n that there are exactly n customers in the shop is $P_0 = 1/16$, $P_1 = 4/16$, $P_2 = 6/16$, $P_3 = 4/16$, $P_4 = 1/16$.

   (*a*) Calculate L_S, queue length.

   (*b*) Determine the expected number of customers being served.

   (*c*) Given that an average of 4 customers per hour arrive and stay to receive a hair cut, determine W_S and W_Q.

9. Explain why the utilization factor ρ for the server in a single-server queuing system must equal $1 = P_0$ where P_0 is the probability of having 0 customer in the system.

10. The jobs to be performed on a particular machine arrive according to a Poisson input process with a mean rate of two per hour. Suppose that the machine breaks down and will require one hour to be repaired. What is the probability that the number of new jobs that will arrive during the time is (*a*) 0, (*b*) 2, and (*c*) 5 or more.

❑❑❑

SYSTEM DYNAMICS

Control model of autopilot aircraft discussed in chapter one shows how various parts and their activities are controlled by gyro, and redirects the aircraft in the desired direction. Can we apply control theory to every day life? It was seen in the example of autopilot that main concern of the control mechanism is to control the stability and oscillations of the system.

There are many examples in nature where we can also apply control theory. Like engineering problems, instability and oscillations also occur in nature. Let us take the example of population explosion. It is observed that if we do not control the population of human being or even any other specie, it will grow exponentially. This is one of the fields in nature, where control theory is applied to study the growth of population. In the field of medicines, multiplication of cancerous cells in human body is another example. Similarly in market there are oscillations in prices of products, which effect their supply and production. In Physics, decay of radioactive material can also be modeled with the help of control theory. Although the precision applicable to engineering problems can not be attained in such problems yet control theory can suggest changes that will improve the performance of such systems.

In scientific literature studies connected with industrial problems are called *Industrial Dynamics* (Geoffrey Gordon), where as study of urban problems is called *Urban Dynamics*. Similarly control of environmental problems is called *World Dynamics*. In all these systems there is no difference in the techniques to be used to study the system, therefore it is appropriate to call this field as System Dynamics.

The principal concern of a system Dynamics study is to understand the forces operating on a system, in order to determine their influence on the stability or growth of the system (Geoffrey Gordon). Output of such study may suggests some reorganization, or changes in the policy, that can solve an existing problem, or guide developments away from potentially dangerous directions. Unlike engineering problems, in this case system dynamics may not produce certain parameters to improve the performance of system. But this study definitely helps the system analyst to predict the scenario so that corrective steps can be taken in time.

8.1 Exponential Growth Models

There are various activities in nature, in which rate of change of an entity is proportional to itself. Such entities grow exponentially and study of such models are known as exponential growth models. To understand this concept, let us consider the birth rate of monkeys. Unlike other animals, birth rate of monkeys grows very fast. If not controlled, their population grows exponentially. Let us try to model this simple natural problem mathematically. If in a region, say x is the number of monkeys at time t, then their rate of growth at time t is proportional to their number x at that time. Let proportionality constant be k. This type of functions can be expressed in the form of differential equations as,

$$\frac{dx}{dt} = kx \qquad ...(8.1)$$

with the condition $x = x_0$ at $t = 0$.

This is first order differential equation and its solution is

$$x = x_0 e^{kt} \qquad ...(8.1a)$$

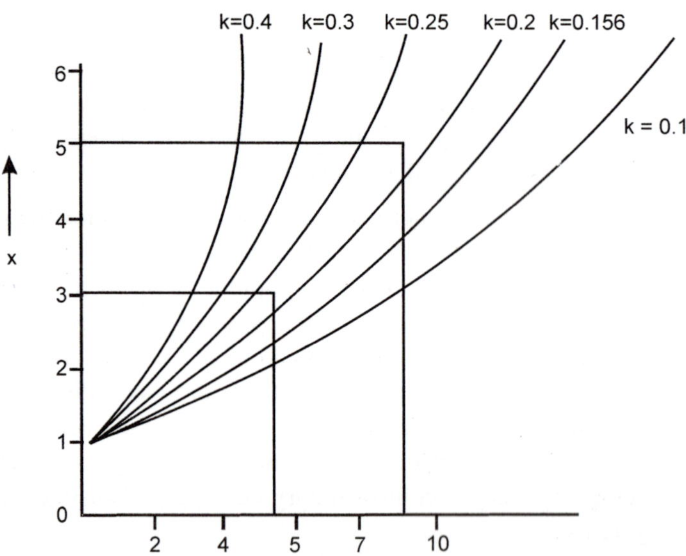

Fig. 8.1: *Exponential growth curves.*

Figure 8.1 shows variation of x vs. t for different values of k. Figure shows, higher is the value of k, more steep is the rise in x. These types of curves are called exponential growth curves and equation (8.1*a*) is a mathematical expression for exponential curve. We now apply this law to the population of monkeys.

Let us take example of population of monkeys in a city. Let their production period is six months. Assuming that in this city, there are 500 monkeys i.e., 250 couples at time $t = 0$. If each couple produces four offspring's in time = 1 (six months), then proportionality constant $k = 2$. In first six months population becomes 1850 (7.4 times) by this relation and in next six months it becomes 13684, which is equal to 54.7 times of 250.

We can also express function (eq. 8.1) on a semi lag graph. Equation (8.1) can be written as

$$\ln x = \ln x_0 + 2t \quad ...(8.1b)$$

which is equation of straight-line in '$\ln x$' (natural log of x) and t. This means that, if we plot this equation by taking x on log axis and t on simple axis, we will get output as a straight-line, with slope equal to 2. What are the dimensions of k? From equation (8.1a) it can be seen that dimensions of k are nothing but 1/time. Sometimes coefficient k is written as $1/T$, that is total period under study. Thus equation (8.1a) can be written as

$$x = x_0 e^{t/T}$$

The constant T is said to be *time constant*, since it provides the measure of growth of x.

8.2 Exponential Decay Models

Radioactive materials continuously decay, because they radiate energy and thus lose mass, and ultimately some part of the matter is changed to some other material. This is proved due to the fact that one can not get pure radium from any ore. It is always be mixed with some impurities such as carbon. It is thought that radium was 100% pure when earth was formed. By finding the quantum of carbon in the radium ore, scientists have determined the life of Earth. Let us see, how this situation is modeled Mathematically. If rate of change of variable x is proportional to its negative value, then such growth is called negative growth and such models are called negative growth models, or exponential decay models. The equation for such a model is

$$\frac{dx}{dt} = -kx \quad ...(8.2)$$

with the condition $x = x_0$ at $t = 0$.

Solution of this equation is

$$x = x_0 e^{-kt} \quad ...(8.2a)$$

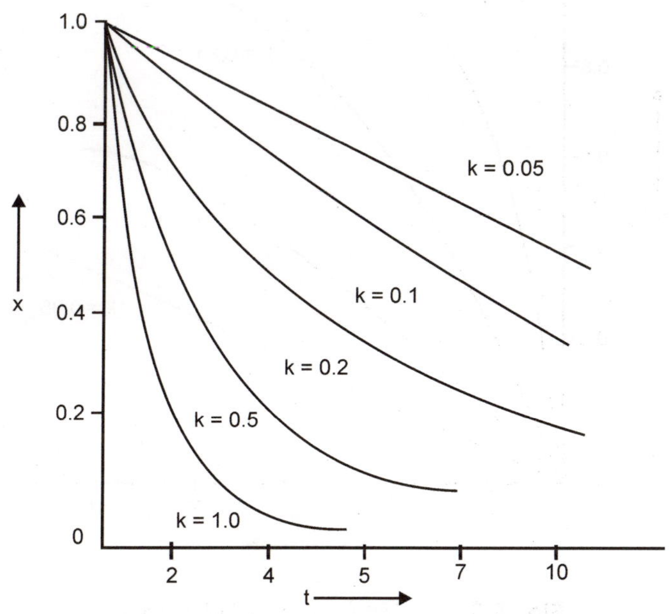

Fig. 8.2: *Exponential decay curve.*

Equation 8.2a is plotted in Figure 8.2. From the figure 8.2 we can see that value of x at any time t, decreases from its initial value and comes down to zero when t is infinity. Decay rate increases with k.

8.3 Modified Exponential Growth Model

A production unit, which is planning to launch a new product, first problem faced by it is, how much quantity of a product can be sold in a given period. A market model should be able to predict, rate of selling of a product, which obviously cannot be proportional to itself. There are several other parameters to be considered while modeling such a situation. In practice, there is a limit to which one can sell the product. It depends on how many other brands are available in market, and what is the probable number of customers. The exponential growth model can not give correct results as it shows unlimited growth. Thus we have to modify this model.

Exponential growth model can be modified if we assume that rate of growth is proportional to number of people who have yet not purchased the product. Suppose the market is limited to some maximum value X, where X is the number of expected buyers. Let x be the number of people who have already bought this product or some other brand of same product. The numbers of people who have yet to buy are $(X - x)$. Thus equation (8.1) can be modified as

$$\frac{dx}{dt} = k(X - x) \qquad \text{...(8.3)}$$

with the condition $x = 0$ at $t = 0$. Solution of equation (8.3) is as

$$x = X(1 - e^{-kt}) \qquad \text{...(8.3}a\text{)}$$

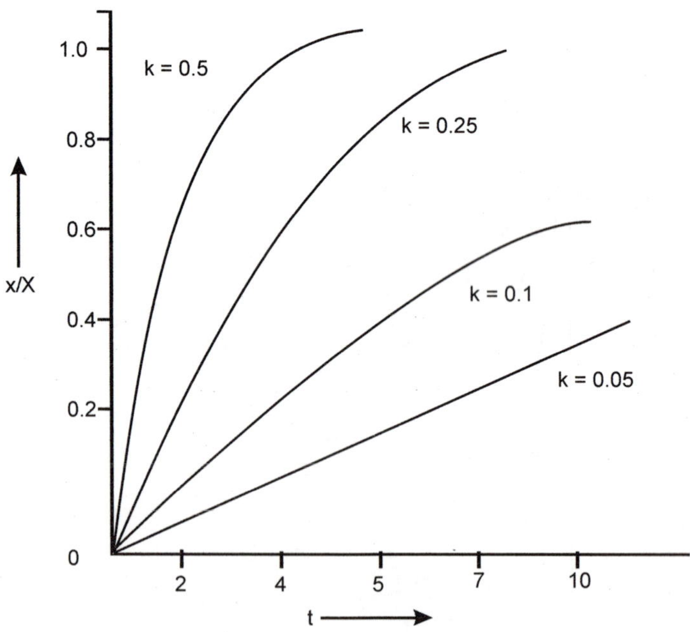

Fig. 8.3: *Modified exponential model.*

Figure 8.3 gives the plot of equation (8.3*a*) for various values of k. This type of curve is sometimes referred as *modified exponential curve.* As it can be seen, maximum slope occurs at the origin and the slope steadily decreases as time increases. As a result of it, the curve approaches the limit more slowly, and never actually gets the limit. In marketing terms, the sale rate drops as the market penetration increases. The constant k plays the same role as the growth rate constant as in Exponential growth model. As k increases, the sale grows more rapidly. As with the growth model, k is sometimes expressed as equal to $1/T$, in which case it can be interpreted as a time constant.

Example 8.1: A builder observes that the rate at which he can sell the houses, depends directly upon the number of families who do not have a house. As the number of families without house diminish, the rate at which he sells the houses drops. How many houses in a year can he sell?

Solution: Let H be the potential number of households and y be the number of families with houses. If $\frac{dy}{dt}$ is the rate at which he can sell the houses, then $\frac{dy}{dt}$ is proportional to $(H - y)$, i.e.,

$$\frac{dy}{dt} = k(H - y), \; y = 0 \text{ at } t = 0$$

This is nothing but Modified Growth case and solution is

$$y = H(1 - e^{-kt})$$

where H is the potential market.

Example 8.2: Radioactive disintegration

The rate of disintegration of a radioactive element is independent of the temperature, pressure, or its state of chemical combination. Each element thus disintegrates at a characteristic rate independent of all external factors. In a radioactive transformation an atom breaks down to give one or more new atoms.

If to start with $t = 0$, the number of atoms of A present is a (say). After time t, x atoms will have decomposed leaving behind $(a - x)$ atoms. If then in a small time interval dt, dx is the number of atoms which change, the rate of disintegration $\frac{dx}{dt}$ can be expressed as

$$\frac{dx}{dt} = k(a - x) \qquad \text{...(8.4)}$$

This is called law of mass action and k is called *velocity constant* or *disintegration constant* or *transformation constant*. Equation (8.4) on integration, with initial condition, $x = 0$ at $t = 0$ gives

$$\ln \frac{a - x}{a} = -kt$$

or

$$a - x = ae^{-kt}$$

If we write $y = (a - x)$ and $a = y_0$, we get

$$y = y_0 e^{-kt}$$

which is same as equation (8.2*a*). If T is the time when half of the element has decayed i.e., $x = a/2$, we get T, as

$$T = \frac{2.303}{k}\log 2$$

$$T = 0.693/k$$

This means the disintegration of an element to half of its period T depends only on k and is independent of amount present at time $t = 0$.

8.4 Logistic Models

Let us again come back to sale of a product in the market. The model of section 8.2 is some what unrealistic because, in modified exponential model, the slope of the product in the beginning is shown to be maximum. In fact in the beginning sale is always less and when product becomes popular in the market, sale increases, that is, slope increases as occurs in the exponential growth model. When for the sale of product, saturation comes, slope again decreases, making the curve of market growth like modified exponential curve. The result is an *S*-shaped curve as shown in Figure 8.4.

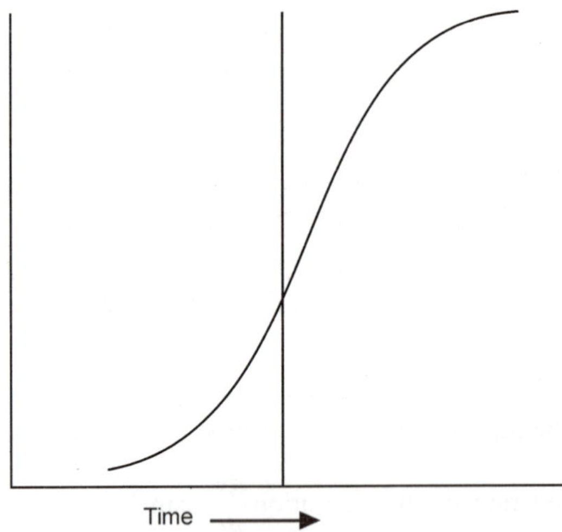

Fig. 8.4: *S-shaped growth curve.*

Such curves are called logistic curves.

The logistic function is, in effect, a combination of the exponential and modified exponential functions, that describes this process mathematically. The differential equation defining the logistic function is

$$\frac{dx}{dt} = kx(X - x) \qquad \text{...(8.5)}$$

In this relation, initially x is very small and can be neglected as compared to X. Thus equation (8.5) becomes

$$\frac{dx}{dt} = kxX$$

which is the equation for exponential growth curve with proportionality constant equal to kX. Much later, when the market is almost saturated, the value of x becomes comparable to X, so that it changes very little with time. The equation for the logistic curve then takes the approximate form

$$\frac{dx}{dt} = kX(X-x)$$

which is the differential equation for the modified exponential function with a constant kX.

The true differential equation is nonlinear and can be integrated numerically with the boundary conditions $x = 0$, when $t = 0$. Exact analytic solution is tedious and has been given by (Croxton et al., 1967). Interested students may see the reference. Apart from market trends, many other systems follow logistic curve, for example population growth can also follow logistic curve (Forrester JW, 1969). During initial stages, there may be ample resources for the growth of population but ultimately when resources reduce to scarcity, rate of population comes down.

The model is also applicable to spread of diseases. Initially it spreads rapidly as many acceptors are available but slowly people uninfected drop and thus growth rate of disease also decreases.

8.5 MULTI-SEGMENT MODELS

In market model, we can introduce more than one products, so that sale of one depends on the sale of other and thus both are mutually related. For example in example 8.1, we have considered the case of a builder who wants to sell houses. It was observed that rate of sale of houses depended on the population which did not have houses. Now suppose another firm wants to launch its product, say air conditioners in the same market. He can only sell air conditioners to people who already have purchased houses from the builders. Otherwise they will not require air conditioners. Let us make a model of this situation. This model can be constructed as follows:

Let at time t, H be the number of possible house holds, y, the number of houses sold and, x the number of air conditioners installed. Then

$$\frac{dy}{dt} = k_1(H-y)$$

$$\frac{dx}{dt} = k_2(y-x) \quad \text{...(8.6)}$$

Equation (8.6) means the rate of sale of houses at time t, is proportional to $(H - y)$, which is the number of people who do not have houses. Similarly second equation of (8.6) says, the rate of sale of air conditioners is proportional to $(y - x)$ i.e., the number of houses which do not have air conditioners installed so far. Both these equations are modified exponential growth models. In equation (8.6), we have not taken into account the air conditioners which become unserviceable and require replacement. This factor can be added by modifying second equation of (8.6) as,

$$\frac{dx}{dt} = k_2(y-x)-k_3x \quad \text{...(8.6}a\text{)}$$

Equations (8.6) and (8.6*a*) can easily be computed numerically, when k_1, k_2 and k_3 are known.

8.6 Modeling of a Chemical Reaction

In a chemical reaction, reaction velocity is very much similar to that of velocity of motion in kinetics and the term denotes *the quantity of a given substance which undergoes change in unit time*. In this process, the rate of reaction is never uniform and falls of with time the reactants are used up. A case of unimolecular reaction has already been discussed in example 8.2 of this chapter. It is known that in a chemical reaction, velocity of reaction $\frac{dx}{dt}$, where x is available reactant is proportional to $(a - x)$ where $x = a$ at the beginning of reaction i.e. at $t = 0$. This is called *law of mass action* in chemistry. That is if x is a molecular concentration of a reactant at any time t and k is the proportionality constant, then

$$\frac{dx}{dt} = (a - x)$$

In a chemical reaction, the number of reacting molecules whose concentration alters as a result of chemical changed as order of reaction.

Consider a general reaction,

$$nA + mB \rightarrow pC + qD$$

where concentration of both A and B alters during the reaction. At any time t, velocity of this chemical reaction is given by,

$$\frac{dx}{dt} = kA^n\ B^m$$

where $(m + n)$ represents the order of the reaction. One example of reaction of first order has already been considered in example 8.2. Let us consider another example below.

Example 8.3: Following data as obtained in a determination of the rate of decomposition of hydrogen peroxide, when equal volumes of the decomposing mixture were titrated against standard $KMnO_4$ solution at regular intervals:

Time (in minutes)	0	10	20	30
Volume of $KMnO_4$	25	16	10.5	7.08

used (in cc's)

Show that it is a unimolecular reaction.

Solution: For a unimolecular reaction,

$$k = \frac{2.303}{t}\log_{10}\frac{a}{a-x}$$

Here the volume of $KMnO_4$ solution used at any time corresponds to undecomposed solution i.e., $(a - x)$ at that time. The initial reading corresponds to a. Inserting experiment values in above equation one gets,

$$k' = \frac{1}{10}\log_{10}\frac{25}{16} = 0.0194$$

$$k' = \frac{1}{20}\log_{10}\frac{25}{10.5} = 0.0188$$

$$k' = k\ \frac{1}{30}\log_{10}\frac{25}{7.08} = 0.0183$$

The constant value of k' shows that the decomposition of $KMnO_4$ is a unimolecular reaction.

8.6.1 Second Order Reaction

In a second order reaction, the minimum number of molecules required for the reaction to proceed is two. Let a be the concentration of each of the reactants to start with and $(a - x)$ their concentration after any time t. Then we have in this case,

$$\frac{dx}{dt} = k(a-x)(a-x)$$

$$= k(a-x)^2 \quad \text{(Law of mass action)}$$

Initial condition for integrating this equation are, at $t = 0, x = 0$. Thus equation becomes on integration,

$$x = \left[\frac{a^2kt}{1+akt}\right]$$

If we start with different amounts a and b of the reactants, then according to law of mass action,

$$\frac{dx}{dt} = k(a-x)(b-x)$$

with initial conditions, at $t = 0$, $x = 0$, one gets,

$$\frac{\left(1-\frac{x}{a}\right)}{\left(1-\frac{x}{b}\right)} = \exp(kt(a-b))$$

8.7 Representation of Time Delay

In all the models so far discussed, we considered different proportionality constants. By dimensional analysis, we observe that these constants have dimensions of 1/time. Thus if inplace of these constants, we take average time to complete a task, it will be more meaningful. Let us consider market model of section 8.4. If average time required to complete the housing project of an area is T_1, say 10 years and time required to install air conditioners in available houses is T_2, then equation (8.6) can be written as

$$\frac{dy}{dt} = \frac{1}{T_1}(H-y)$$

$$\frac{dx}{dt} = \frac{1}{T_2}(y-x) \quad \text{...(8.7)}$$

This is simplistic form of model. Determination of T_1 and T_2 is not that straight forward. For example T_1 depends on number of factors viz., economic conditions of the people, cost of land, speed of housing loans available etc. Similarly, T_2 depends on again, how many people owning house, can afford air conditioners, and also it depends on weather conditions and many other factors. In order to model such situations, one has to take all the factors into account. In order to model such situation correctly, feedback of information regarding market trends, is very much essential. Without this information, air conditioner vendor, may stock air conditioners based on houses available. If he is not able to install the air conditioners because of other conditions, he will have to stock them for longer time and bear losses. This becomes another problem called inventory control. Thus all the parameters have to be taken into account while constructing such models.

8.8 A Biological Model

Here we consider a biological model, an application of System Dynamics. There are many examples in nature of parasites, that must reproduce by infesting some host animal and, in doing so, kill the animal. As a result, the population of both, the host and parasite fluctuate. As the parasite population grows, host population declines. Ultimately, decline in host population results in the decline in parasite population, owing to which host population starts increasing. This process can continue to cause oscillations indefinitely.

To construct this model of balance between host and parasite, let x be the number host and y be the number of parasites at a given time t. Let birth rate of host over the death rate due to natural causes is a, where a is positive. In the absence of parasites, the population of hosts should grow as,

$$\frac{dx}{dt} = \alpha x$$

The death rate from infection by the parasites depends upon the number of encounters between the parasites and hosts, which is assumed to be proportional to the products of the numbers of parasites and hosts at that time. Thus the rate of growth of hosts modifies to,

$$\frac{dx}{dt} = \alpha x - kxy \qquad \text{...(8.8)}$$

Here a simplifying assumption, that each death of a host due to parasite results in the birth of a parasite. This is the only mean, by which parasite population can grow. It is also assumed that death rate of parasites is δ due to natural death. Thus the equation controlling the parasite population is,

$$\frac{dy}{dt} = kxy - \delta y \qquad \text{...(8.9)}$$

We can solve equations (8.8) and (8.9) numerically by taking values of parameters α, k, and δ as, $\alpha = 0.005$, $k = 6 \times 10^{-6}$, and $\delta = 0.05$.

Table 8.1

Day no.	Host population	Parasite population
0	10000	1000
100	7551.74	1303.45
200	7081.92	570.282
300	8794.23	446.034

Contd...

Day no.	Host population	Parasite population
400	10021.9	967.877
500	7634.68	1327.57
600	7047.85	585.517
700	8722.74	441.226
800	10037.3	936.287
900	7721.91	1349.8
1000	7016.39	601.652

Solution: If we assume that initial value of host and parasite is $x_0 = 10000$ and $y_0 = 1000$, then following steps will compute the population, by taking t as one day.

$$x_{i+1} = x_i + (\alpha x_i - kx_i y_i)dt$$

$$y_{i+1} = y_i + (kx_i y_i - \delta y_i)dt$$

where $i = 0, 1, 2, 3,\ldots$

Results of computation is given in Table 8.1.

Example 8.4: Babies are born at the rate of one baby per annum for every 20 adults. After a delay of 6 years, they reach school age. Their education takes 10 years, after which they are adults. Adults die after an average age of 50 years. Draw a system dynamic diagram of the population and program the model, assuming the initial number of babies, school going children and adults are respectively, 300, 3000, and 100,000.

Solution: Let at time t, there are x babies and y adults in the population. Then after time period δt, increase in number of babies will be,

$$\delta x = \frac{z}{20}\cdot \delta t - \frac{x}{6}\delta t$$

On dividing by δt and as $\delta t \to 0$, one gets

$$\frac{dx}{dt} = z/20 - x/6$$

By similar logic, if y is the population of school going children at time t, then

$$\frac{dy}{dt} = \frac{x}{6} - \frac{y}{10}$$

and if z is the population of adults at time t, then

$$\frac{dz}{dt} = kxy - \delta y$$

The above equations can be solved numerically by taking initial values for x, y, and z as 300, 3000, and 100,000. Time period t can be taken as one year. Result of this computation is given below.

Year	Babies	Children	Adult	Total population
0	300	3000	100000	103300
1	5250	2750	98300	106300
2	9290	3350	96700	109340
3	12577	4563	95101	112241

EXERCISE

1. In the model of house contractor and air conditioners (section 8.5), assume that average time to sell a house is 4 months, average time to install an air conditioner is 9 months, and break down of air conditioner occurs, on average 25 month. Take the initial housing market to be 1000 houses. Assume defected air conditioners to be replaced. Then in a span of 5 years, how many houses and air conditioners will be sold.
2. In which type of applications, control models are used.
3. Derive an expression for exponential decay models and give one example where it is used.
4. If at time $t = 0$, radioactive material in a compound is a and proportionality constant is k, find an expression for half-life of the radioactive part.
5. With reference to market model, is the modified exponential model realistic? Explain in details, if not, how this model is modified.
6. Give a mathematical model of logistic curve.
7. For a species of animals in an area, the excess of the births over natural deaths causes a growth rate of a times the current number N. Competition for food causes deaths from starvation at rate bN^2. Simulate the population growth assuming $a = 0.05$, $b = 0.00001$, and $n = 1000$ at time $t = 0$.
8. A certain radio-element has disintegration constant of 16.5×10^6 sec^{-1}. Calculate its half-life period and average life.

9

INVENTORY CONTROL MODELS

In chapter seven, we have studied the application of simulation of modeling in various systems where queuing is involved. Simulation and modeling has application in almost all the branches of science, especially, where events are stochastic in nature. Inventory control is one such field and will be studied in this chapter. Whether it is a manufacturing unit or a sale outlet, one of the pressing problem faced by the management is the control of inventory. Many companies fail each year due to the lack of adequate control of inventory in their stores. Whether it is raw material used for manufacturing a product or products waiting for sale, problem arises when, too few or too many items are stored in the inventory. If the number of items stored are more than what are required, it is a loss of investment and wastage of storage space which again results in the loss of investments. In the case of excess inventory, items may depreciate, deteriorate, or become obsolete. But if less number of items are kept in store, it can result in the loss of sale or reduction in the rate of production, which ultimately results in the loss of business. In this case there will be loss of profit because of loss of sale and loss of goodwill due to unfilled demand. Also stock has to be replenished frequently which involves replenishment cost. Then question arises, how much to store in the inventory at a given span of time. This in turn depends on what is the annual demand and how much time it takes to replenish the inventory by the supplier. There can be uncertainties, such as strike, weather calamities, price hikes and so on, in replenishment of inventory. While computing the inventory for storage, all these factors are to be taken into account. Thus basic problem in inventory control is to optimize the sum of the costs involved in maintaining an inventory. Backordering is the case when inventory goes to zero and orders are received for the sale of item, or raw material is required for production. When fresh inventory arrives, first back orders are completed and then regular orders are entertained. In this case, raw material is required for production purpose and inventory goes to zero, production stops, which is a loss to firm. But if inventory is of items, which are for the sale, goodwill of customer is lost in this case and even there is a possibility that customer will also be lost. Thus backordering case should always be avoided. In the next section we will discuss basic laws of inventory control models and also make simulation models for some case studies.

The mathematical inventory models used with this approach can be divided into two broad categories—deterministic models and stochastic models, according to the predictability of demand involved. If the demand in future period can be predicted with considerable precision, it is reasonable to use an inventory policy that assumes that all forecastes will always be completely accurate. This is the case of known demand where a *deterministic inventory* model would be used. However when demand can not be predicted very well, then in this case *stochastic inventory* model will be required to be used, where demand in any period is a random variable rather than a known constant. These cases are discussed in following sections.

9.1 Infinite Delivery Rate with No Backordering

In the present section, we take the simplest case when there is no back ordering and demand is deterministic. Case of infinite delivery with no backordering will be first modelled and then extended to other cases. Costs involved in maintaining an inventory in this case are,

c_0 = costs associated with placing an order or setting up for a production run i.e., ordering cost.

c_1 = inventory holding cost (sometimes called *storage costs*).

The key to minimizing inventory costs is for deciding, when to order, how to order, and how much to order, and how much back ordering to allow (if applicable). If the demand is known and the time to receive an order (lead time) is constant then when to order is not a problem. We make following assumptions while modeling the inventory problem.

1. A single type of product is analyzed even though many types are held in inventory for use or sale.
2. The planning period is 1 year.
3. The demand for the product is known and constant throughout the year.
4. The lead time is known and is constant (time between the order and receipt of the order).
5. Complete orders are delivered at one time (infinite delivery rate).
6. Unfilled orders are lost sales (no backordering allowed).

Thus only two costs are involved in this model i.e., ordering cost and inventory holding cost.

Generally multiple types of products are kept in the inventory and their costs are also interdependant. But in this section we will consider only single product and model valid for single product can easily be extended for case of multiple products. Generally two types of products are kept in the inventory, for manufacturing of some item and for direct sale. Inventory which is stocked for the use in manufacturing will be called raw materials. If management allows the stock of raw material to be depleted, several conditions may arise i.e.,

- Substituted or borrowed raw material will be used.
- Emergency measures may be taken for the supply of raw material.
- The company may switch to the manufacturing of different product.
- The manufacturing process may shut down completely.

If management allows product to be depleted, loss of profit will result. In case of depletion of material for sale, firm may lose the customer permanently.

Here it is assumed that *lead time*, that is time gap between the requisition and receipt of inventory, is constant. Let us define some parameters related with inventory as,

a = rate of depletion of inventory,

c = cost of purchase of one unit,

h = holding cost per unit per unit of time held in inventory,

Q = quantity which is ordered each time (Economic Order Quantity (EOQ)),

D = annual demand for the product,

c_0 = set-up cost involved per order (ordering cost),

$Q/2$ = annual inventory level,

N = D/Q = number of orders per year,

AIHC = annual inventory holding cost.

Thus Annual Ordering Cost (AOC) is given by

$$\text{AOC} = c_0 \cdot (D/Q) \quad \text{...(9.1)}$$

It is to be noted that c_0 is the cost per order, and does not include the cost of material to be ordered. Annual inventory holding cost is given as,

$$\text{AIHC} = h \cdot (Q/2) \cdot (Q/a) = hQ^2/2a \quad \text{...(9.2)}$$

where Q/a is the time period after which order is to be placed.

Therefore total inventory cost per cycle is given by

$$\text{Total cost per cycle} = c_0 + hQ^2/2a + c \cdot Q \quad \text{...(9.3)}$$

which is sum of ordering, holding and material costs. Therefore total cost per unit time is,

$$T = c_0 a/Q + hQ/2 + c \cdot a$$

Our aim is to minimise T. For T to be minimum, we use the concept of maxima-minima theorem, which says, for a function to be minimum, its first derivative is equal to zero whereas second derivative is positive, Thus,

$$\frac{dT}{dQ} = 0$$

and

$$\frac{d^2T}{dQ^2} > 0 \quad \text{...(9.4)}$$

First of equation (9.4) gives,

$$-\frac{ac_0}{Q^2} + h/2 = 0 \quad \text{...(9.5)}$$

or

$$Q = \sqrt{2ac_0/h}$$

Second of equation (9.4) gives

$$\frac{d^2T}{dQ^2} = 2ac_0/Q^3 > 0$$

Thus value of Q given in equation (9.5) gives the minimum value of T.

Equation (9.5) is the well known EOQ formula [21]. Figure 9.1 gives the graphic representation of inventory depletion and replenishment. The corresponding cycle time t, is given by,

$$Q/a = \sqrt{2c_0/ah}$$

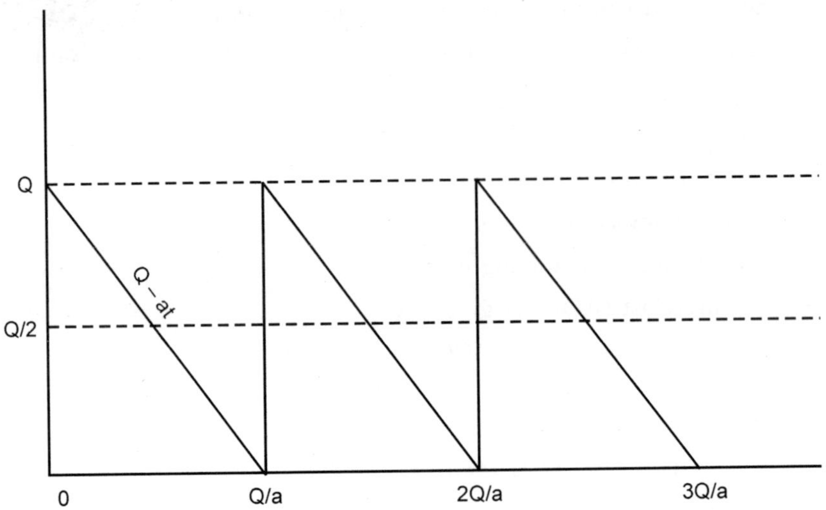

Fig. 9.1: *Diagram of inventory level as a function of time for EOQ model.*

Example 9.1: Calculate the EOQ from the following:

Annual requirement = 50 units/month

Ordering cost/order = Rs.10.00

Material cost/unit = Rs.6.00

Inventory holding cost/unit = 20% of unit cost.

Solution: We take unit of time as one year. Total annual requirement is 600 units/year, which is equal to a. Cost c_0 is Rs.10.00/unit. Holding cost h is Rs.1.20/unit/time, thus,

$$\text{EOQ} = Q = \sqrt{2ac_0/h} = \sqrt{2 \times 600 \times 10/1.20} = 100 \text{ units.}$$

Example 9.2: The demand for a particular item is 18000 units per year. The holding cost per unit is Rs.1.20 per year and the cost of one procurement is Rs.400.00. No shortage is allowed, and the replacement rate is instantaneous. Determine;

(*a*) Optimum order quantity,

(*b*) Number of orders per year,

(*c*) Time between orders

(*d*) Total cost per year when the cost of one unit is Rs.1.00.

Solution: Here a = 18000 units per year, h = Rs.1.20 /year, c_0 = Rs.400/order. We take one year as the unit of time.

(*a*) Optimum order quantity $Q = \sqrt{2ac_0/h} = \sqrt{2 \times 18000 \times 400/1.20}$ = 3464.10 units = 3465 units.

(*b*) Number of orders per year = total time period/cycle time = a/Q = 18000/3465 = 5.2 or 5 orders/year.

(*c*) Time between orders = 365/5 = 73 days

(*d*) Total cost = material cost + storage cost + ordering cost

= unit price × no. of units + $(Q/2)h$ + $(ac_0)/Q$

= 1 × 18000 + 3465 × 1.20/2 + 18000 × 400/3465

= Rs.22157.00

9.2 Finite Delivery Rate with No Backordering

Suppose there is a case, when complete order is not deliverd in one instalment, but is sent in part deliveries. This may be due to nonavailability of raw material, or order is so bulky that it can not be carried in single instalment. We also assume that arrival rate is greater than the use or sail rate, thus there is no depletion. Such case is called finite delivery rate with no back ordering, and is modelled as follows.

Let A = arrival rate of an order in units/day

a = use or sale rate in units/day

Q = order quantity

Q/A = time to receive complete order of Q units (in days)

and $A > a$, which means, arrival rate is greater than sale rate.

Since units are sold or used while others are arriving and being added to the inventory, the inventory level will never reach the EOQ level. Since arrival rate is A/day, total ordered quantity is supplied in Q/A days. Now a number of units are used in one day, therefore aQ/A is the number of units sold in Q/A days, when whole order is supplied. Then since $A > a$, there are always some items left in inventory, thus

$$Q - a\frac{Q}{A} = \text{maximum inventory level in one cycle } (Q/A \text{ days}).$$

$$\frac{1}{2}[Q - a\frac{Q}{A}] = \text{average inventory level in one cycle.}$$

Now total inventory cost = holding cost + annual ordering cost + material cost, then total cost per unit time t, is

$$T = \frac{h}{2}[Q - a\frac{Q}{A}] + \frac{c_0 a}{Q} + ca \qquad \text{...(9.6)}$$

where c_0 is the cost involved per order and a/Q is number of orders per unit time, h being the cost of holding inventory per unit time. Differentiating this equation with respect to Q and equating the resulting equation to zero as in section 9.1, we get

$$Q = \sqrt{\frac{2c_0 a}{h \cdot (1 - a/A)}} \qquad \text{...(9.7)}$$

which is the expression for optimum quantity to be ordered, when arrival rate and depletion rate are known.

Example 9.3: A small manufacturing company specialises in the production of sleeping bags. Based on the past records, it is estimated that the company will be able to produce 5000 bags during the next year if the raw materials is available, when needed. Raw material for each bag costs Rs.50.00.

Assuming that bags are produced at constant rate during the year of 300 working days, it is estimated that the annual holding cost of the inventory of raw material is 20% of the raw material cost. Also each time order is placed, company has to pay Rs.25.00 as reordering cost. If lead time is 7 days, calculate total annual inventory cost, and total cost.

Solution: Since manufacturer is planning to manufacture 5000 bags in next year, this means sale (depletion) rate per year is 5000. If an year is taken as time unit, then a = 5000, c_0= 25.00, h = 0.2 × 50 = 10.00, therefore

$$\text{TAIC}(Q) = c_0 \cdot \frac{a}{Q} + h \cdot \frac{Q}{2}$$

$$\text{Now } Q = \sqrt{2ac_0/h}$$

Therefore,

$$\text{TAIC}(Q) = c_0 \cdot \frac{a}{Q} + h \cdot \frac{Q}{2} = c_0 \frac{a}{\sqrt{2ac_0/h}} + h \frac{\sqrt{2ac_0/h}}{2}$$

$$= \sqrt{c_0 \cdot h \cdot a/2} + \sqrt{\frac{2a \cdot c_0 \cdot h}{4}}$$

$$= \sqrt{25 \times 10 \times 5000/2} + \sqrt{50 \times 10 \times 5000/4}$$

$$= 790.57 + 790.57 = 1581.14$$

where $Q = \sqrt{2ac_0/h} = \sqrt{2 \times 5000 \times 25/10}$

= 158.11 order-quantity

Therefore total cost (TC) = cost of the material + TAIC

= 50 × 5000 + 1581.14 = 251581.14

This is the long-run average total cost since

$$N = a/Q = 5000/158 = 31.65 \text{ orders/year}$$

Since N is not an integer, 31 orders will be placed in one year and 32 in the next year. Production per day is 5000/300 = 16.66, and lead time for supply of raw material is 7 days. Replenishment order should go, when raw material is left for only 16.66 × 7 = 117 i.e., for 117 bags.

Example 9.4: In example 9.3, if supplier gives 5% discount to manufacturer, on the condition that, he purchases material, only twice in a year. Is this proposal acceptable?

Solution: In this case cost per bag is

$$\text{Cost/bag} = \text{Rs.}(50.00 - 0.05 \times 50) = \text{Rs.}47.50$$

$$Q = 2500$$

$$a = \text{annual requirement} = 5000,$$

Thus total cost under this proposal is

$$\text{TC} = \begin{pmatrix}\text{cost of materials} \\ \text{for 5000 bags}\end{pmatrix} + \begin{pmatrix}\text{annual} \\ \text{order cost}\end{pmatrix} + \begin{pmatrix}\text{annual inventory} \\ \text{holding cost}\end{pmatrix}$$

$$= 5000 \times (47.5) + \frac{25 \times 5000}{2500} + \frac{10 \times 2500}{2}$$

$$= 237500.00 + 50.00 + 12500.00 = \text{Rs.}250050.00$$

This cost is less than TC in first case, which is Rs.251581.14, hence proposal is acceptable.

Example 9.5: In example 9.3, if we put one more condition i.e., arrival rate is 30/day then,

$$A = 30/\text{day}$$

and

$$a = 5000/300 = 17/\text{day}.$$

Then

$$Q = \sqrt{\frac{2 \times 25 \times 5000}{10 \times (1 - 17/30)}}$$

$$= \sqrt{\frac{250000}{4.4}} = \sqrt{56818.18} = 238.36$$

Suppose we accept the policy of 238 bags each time an order is placed then

$$\text{TC} = 5000 \times (50) + c_0 a/Q + \frac{hQ}{2}\left(1 - \frac{a}{A}\right)$$

$$= \text{Rs.}250000 + 25 \times 50000/238 + 238 \times 0.4333$$

$$= \text{Rs.}250000 + 525.21 + 515.627$$

$$= \text{Rs.}251040.83$$

which is slightly less than the total cost in example 9.3, which is 251581.14, hence is economical.

9.3 Infinite Delivery Rate with Backordering

In the previous sections, we assumed that unfilled demands are lost demands. However, unfilled demands for salable items do not always result in lost sales. Quite often customers will wait for an ordered item to arrive, or they will permit the merchant to place an order for the item of interest. An unfilled demand, that can be filled at a later date, is known as a *backorder*. As might be expected, there is generally a cost associated with backorders. Clearly, if all demands could be backordered, and if there were no cost for backorders, there would have been no need for inventory. The merchant could take orders and then wait until the most economical time for him to place an order for future delivery. But this is rarely the case. If an item is backordered, there are generally costs associated with it. If the cost of holding the inventory is high relative to these shortage costs, then lowering the average inventory level by permitting occasional brief shortage may be a sound business decision. Costs due to shortage are,

- Loss of good will
- Repeated delays in delivery
- Additional book keeping, loss of cash that would have been available for immediate use.

The basic problem then is, to decide how much order to be placed and how many backorders to allow before receiving a new shipment. In order to model this situation, we, in this section make following assumptions,

1. Backorders are allowed.
2. Complete orders are delivered at one time.
3. All assumptions from Section 9.1 hold.

Then let

Q = ordered quantity,
D = annual demand for the product,
B = number of backorders allowed before replenishing inventory,
c_0 = ordering cost/order,
h = annual inventory holding cost/unit/time,
p = annual backorder cost/unit/time,
t_1 = time for the receipt of an order until the inventory level is again zero,
t_2 = time from a zero inventory level until a new order is received,
t_3 = time between consecutive orders,
N = D/Q = number of orders/year,
$Q - B$ = inventory level just after a batch of Q units is added.

Since a is the rate of depletion of inventory, during each cycle, the inventory level is positive for a time $(Q - B)/a$. The average inventory level during this time is $(Q - B)/2$ units, and corresponding cost is $h(Q - B)/2$ per unit time.

$$\text{Hence holding cost per cycle} = \frac{h(Q-B)}{2} \cdot \frac{(Q-B)}{a} = \frac{h(Q-B)^2}{2a} \quad \text{...(9.8)}$$

Similarly shortage occurs for a time B/a. The average amount of shortage during this time is $(0 + B)/2$ units, and the corresponding cost is $pB/2$ per unit time, where p is the cost per unit short inventory per unit time short.

$$\text{Hence shortage cost per cycle} = \frac{pB}{2} \cdot \frac{B}{a} = \frac{p(B)^2}{2a} \quad \text{...(9.9)}$$

Reordering cost/cycle = c_0, and cost of inventory is cQ. Thus total cost/cycle (TC) is sum of these costs i.e.,

$$\text{TC} = c_0 + \frac{h(Q - B)^2}{2a} + \frac{pB^2}{2a} + cQ \quad \text{...(9.10)}$$

And the total cost per unit time is

$$T = \frac{c_0 a}{Q} + \frac{h(Q-B)^2}{2Q} + \frac{pB^2}{2Q} + ca \quad \text{...(9.11)}$$

Figure 9.2 is the graphic representation of the model.

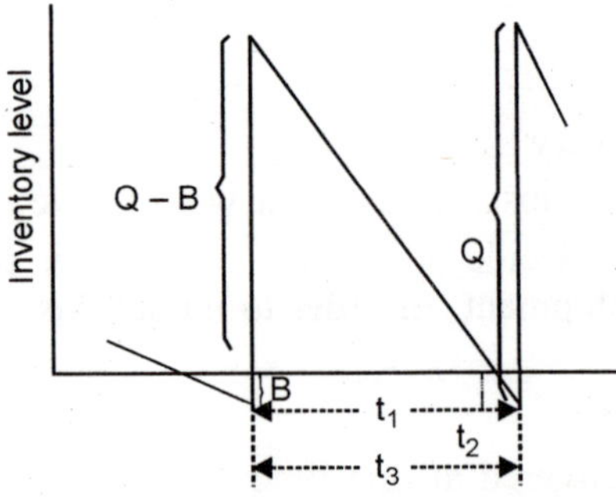

Fig. 9.2: *Graphic representation of backordering model.*

In this model, there are two decision variables (B and Q), so in order to find optimal values of Q and B, we set partial derivatives $\frac{\partial T}{\partial Q}$ and $\frac{\partial T}{\partial B}$ to zero, and solve for Q and B. We get after differentiation,

$$\frac{\partial T}{\partial B} = \frac{-h(Q-B)}{Q} + \frac{pB}{Q} = 0$$

$$\Rightarrow \quad B = \frac{h}{p+h} Q$$

and

$$\frac{\partial T}{\partial Q} = \frac{-c_0 a}{Q^2} + \frac{h(Q-B)}{Q} - \frac{h(Q-B)^2}{2Q^2} - \frac{pB^2}{2} = 0$$

$$\Rightarrow \quad 2c_0 a = Q^2\left[\frac{2hp}{p+h} - h\left(\frac{p}{p+h}\right)^2 - p\left(\frac{h}{p+h}\right)^2\right]$$

$$\Rightarrow \quad 2c_0 a = Q^2\left(\frac{p}{p+h}\right)\left[2h - h\left(\frac{p}{p+h}\right) - \left(\frac{h^2}{p+h}\right)\right]$$

$$\Rightarrow \quad 2c_0 a\left(\frac{p+h}{p}\right) = Q^2\left[2h - h\left(\frac{p}{p+h}\right) - \left(\frac{h^2}{p+h}\right)\right]$$

or

$$Q = \sqrt{\frac{2ac_0}{h}}\sqrt{\frac{h+p}{p}}$$

and

$$B = \sqrt{\frac{2ac_0}{p}}\sqrt{\frac{h}{p+h}} \qquad \text{...(9.12)}$$

which is optimum values of Q and B.

Example 9.6: Suppose a retailer has the following information available:

a = 350 units/year

c_0 = Rs.50 per order

h = Rs.13.75 per unit/time

p = Rs.25 per unit/time

LT = 5 days

To minimize the total annual inventory cost when backordering is allowed, how many units should be ordered each time an order is placed, and how many backorders should be allowed?

Solution: In this case, optimum order and back order are,

$$Q = \sqrt{\frac{(2c_0 a)}{h} \times \frac{(h+p)}{p}} = \sqrt{\frac{(2\times 50)}{13.75} \times \frac{13.75+25}{25}} = 63 \text{ units}$$

$$B = \frac{13.75}{25+13.75} \times 63 = 22 \text{ units}$$

Thus, the optimal policy is to allow approximately 22 backorders before replenishing the inventory with approximately 63 units.

9.4 FINITE DELIVERY RATE WITH BACKORDERING

This section is the same as Section 9.3 except we assume that each order for more stock arrives in partial orders at a constant rate each day until the complete order is received. Thus, all assumptions from Section 9.3 hold with the exception of the one just stated. Let

A = arrival (delivery) rate/day,

a = use or sale rate/day,

Q = order quantity,

B = number of backorders allowed before replenishing inventory,

c_0 = ordering cost/order,

h = annual inventory holding cost/unit/time,

c_2 = annual backorder cost/unit,

p = c_2/time,

t_1 = time from zero inventory until the complete order is received,

t_2 = time from receipt of complete order until the inventory level reaches zero again,

t_3 = time from when backordering starts coming in,

t_4 = time from when a new order starts coming in until all backorders are filled (inventory level comes back to zero again),

t_5 = $t_1 + t_2 + t_3 + t_4$,

N = D/Q = number of orders per year.

Based on the above notations, annual order cost is given as

$$\text{AOC}(Q) = \frac{c_0 D}{Q} \qquad \text{...(9.13)}$$

Annual Inventory Holding Cost (AIHC), Annual Backordering Cost (ABC), are given by

$$\text{AIHC}(Q, B) = h\frac{t_1 + t_2}{t_5} \times \frac{Q - t_1 a - B}{2} \qquad \text{...(9.14)}$$

$$\text{ABC}(Q) = c_2 \; \frac{t_3 + t_4}{t_5} \cdot \frac{B}{2} \qquad \text{...(9.15)}$$

In equation (9.14), $(t_1 + t_2)/t_5$ is the ratio of time, for which inventory remains positive. Now t_5 is the total time in which inventory Q has arrived and depleted. Since arrival rate A is greater than the depletion rate a, $t_5 = Q/A$. Similarly from Figure 9.3, $(t_1 + t_2)$ time is the time for which inventory remained positive, while items were arriving and were being consumed simultaneously. This means,

$$t_1 + t_2 = \frac{Q - t_1 a - B}{(A - a)}$$

Now t_1 is the time in which full EOQ i.e., Q will arrive i.e., $t_1 = Q/A$. Substituting expressions for t_1, t_2, t_3, t_4, t_5 in equations (9.13), (9.14) and (9.15) and adding the three costs, one gets

$$\text{TAIC}(Q,B) = \frac{c_0 D}{Q} + \frac{hA}{2Q(A-a)} \cdot \left(Q\frac{A-a}{A} - B\right)^2 + \frac{pAB^2}{2Q(A-a)} \qquad \text{...(9.16)}$$

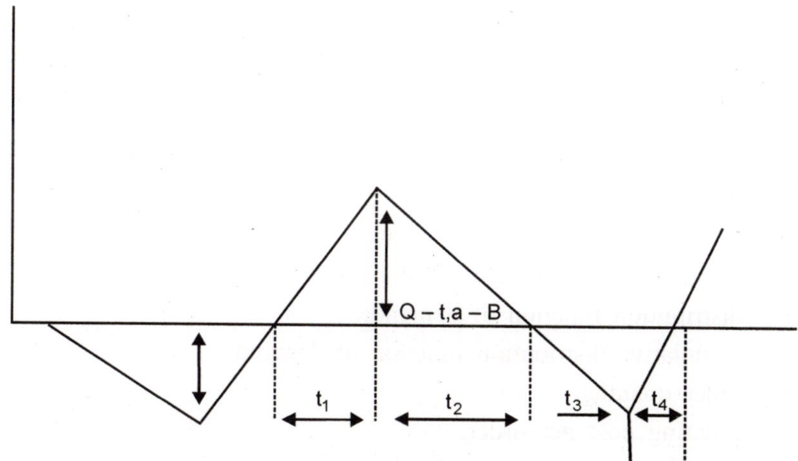

Fig. 9.3: *Finite delivery rate with backordering.*

To find optimum values of Q and B for minimum total annual inventory cost, we differentiate equation (9.16) with respect to Q and B. After differentiation we get after some algebraic computation,

$$Q = \sqrt{\frac{2c_o a}{h(1-\frac{a}{A})} \cdot \frac{h+p}{p}} \qquad \text{...(9.17)}$$

and

$$B = \frac{h}{h+p}\left(1-\frac{a}{A}\right)Q \qquad \text{...(9.18)}$$

9.5 Probabilistic Inventory Models

So far we have discussed the inventory models in which the demand for a single product and the lead time to replenish the inventory were known constants. Consequently, the analysis was straight-forward. Now we construct models where demand and lead time both are not known but are probabilistic. Let us first consider a single-period model.

In the *single period* model, the problem is to determine how much of a single product to have in hand at the beginning of a single time period to minimize the total purchase cost, stockout cost, and ending inventory holding cost. When the demand is random, problem reduces to minimizing the *total expected* inventory cost.

9.5.1 Single-period Models

A single period model holds for the inventory, which is purchased once only as per the season or purchase of perishable items. In each case, the demand for the product is considered to be random variable with a distribution function that is known or can be approximated. The problem is to determine, how much of the product to have on hand at the beginning of the period to minimize the sum of the

- Cost to purchase or produce enough of the product to bring the inventory upto a certain level
- Cost of stock-outs (unfilled demands)
- Cost of holding ending inventory
- Since the demand for the product is a random variable, the number of stockouts encountered and the number of units in ending inventory are also random variables, since they are both function of demand. Hence, the total inventory cost associated with starting the period with a given inventory level is a random variable. In this light, we can only hope to determine the starting inventory level that will minimize the expected value of the three costs that make up the total inventory cost.

Consider the notation to be used in this section:

X = demand for the product during the given period,
$f(x)$ = distribution function of demand,
$F(x)$ = cumulative distribution function of demand,
Q = order quantity,
c_0 = ordering cost per order,
c_1 = inventory holding cost/unit of ending inventory,
c_3 = cost per item purchased or produced,
c_4 = stockout cost/item out of stock,
DIL = desired inventory level at the start of the period,
IOH = initial inventory on hand before placing the order,
TIC(X, DIL) = total inventory cost as a function of demand and the desired inventory level,
TIC(X, DIL, c_0) = total inventory cost as a function of demand and the desired inventory level, and set-up cost c_0.

In the next section, we will take the expected value of the total inventory cost with respect to the demand X, and then determine the value of DIL that will minimize the expected total inventory cost. We will also assume that backordering is permitted, the delivery rate is finite, and lead time is zero.

9.5.2 Single-period Model with Zero Ordering Cost

In this section we assume that there is no ordering cost, and distribution of demand is continuous. Thus total inventory cost is given by,

$$\text{TIC}(X, \text{DIL}) = \begin{cases} c_3(\text{DIL} - \text{IOH}) + c_1(\text{DIL} - X), & X < \text{DIL} \\ c_3(\text{DIL} - \text{IOH}) + c_4(X - \text{DIL}), & X \geq \text{DIL} \end{cases} \qquad \text{...(9.19)}$$

Equation (9.19) states, that if the demand during the period turns out to be less than the inventory level at the start of the period (X < DIL) then total inventory cost is cost of order c_3 (DIL – IOH) plus the cost of holding one unit of inventory time the number of items that will be in the inventory at the end of the period. On the other hand, if the demand is greater than or equal to the desired inventory level, the total inventory cost consists of the cost to order (DIL – IOH) items, plus the cost of a stock-out, times the total number of stock-outs.

The expected total inventory cost can be expressed as

$$E[\text{TIC}(X, \text{DIL})] = \int_{-\infty}^{\infty} \text{TIC}(x, \text{DIL}) f(x)\, dx$$

$$= \int_{-\infty}^{\text{DIL}} [c_3(\text{DIL} - \text{IOH}) + c_1(\text{DIL} - x)] f(x)\, dx$$

$$+ \int_{\text{DIL}}^{\infty} [c_3(\text{DIL} - \text{IOH}) + c_4(x - \text{DIL})] f(x)\, dx$$

$$= c_3(\text{DIL} - \text{IOH}) \int_{-\infty}^{\infty} f(x)\, dx + c_1 \int_{-\infty}^{\text{DIL}} [(\text{DIL} - x)]\, f(x)\, dx$$

$$+ c_4 \int_{\text{DIL}}^{\infty} [(x - \text{DIL})] f(x)\, dx \qquad ...(9.20)$$

$$= c_3(\text{DIL} - \text{IOH}) + c_1 \int_{-\infty}^{\text{DIL}} [(\text{DIL} - x)] f(x)\, dx$$

$$+ c_4 \int_{\text{DIL}}^{\infty} [(x - \text{DIL})] f(x)\, dx$$

which is nothing but sum of cost of Q items, expected holding cost and expected stock-out cost.

Since the expected total inventory cost is not a function of demand, one can take its derivative with respect to DIL, set it equal to zero, and solve for the optimul DIL i.e.,

$$\frac{dE[\text{TIC}(X, \text{DIL})]}{d(\text{DIL})} = c_3 + c_1 \int_{-\infty}^{\text{DIL}} f(x)\, dx - c_4 \int_{\text{DIL}}^{\infty} f(x)\, dx = 0 \qquad ...(9.21)$$

Thus,

$$c_3 + c_1 F(\text{DIL}) - c_4[1 - F(\text{DIL})] = 0$$

$$(c_1 + c_4) F(\text{DIL}) = c_4 - c_3$$

$$F(\text{DIL}) = \frac{c_4 - c_3}{(c_1 + c_4)}$$

The inventory level that will minimize the expected total inventory cost is the value of DIL such that

$$P(X \le \text{DIL}) = F(\text{DIL}) = \frac{c_4 - c_3}{c_1 + c_4} \qquad ...(9.22)$$

The equation (9.22) says $F(\text{DIL})$ represents probability of no stock-outs when the given product is stocked at the optimum DIL. Likewise

$$P(X > \text{DIL}) = 1 - F(\text{DIL}) = 1 - \frac{c_4 - c_3}{c_1 + c_4} \qquad ...(9.23)$$

represents the probability of atleast one stockout (demand X exceeds DIL).

Example 9.7: An outdoor equipment shop in Shimla is interested in determining how many pairs of touring skis should be in stock in the beginning of the skiing season. Assume reordering can not be done because of the long delay in delivery. Last season was a light year, so the store still has 10 pairs of skis on hand. If

(*i*) The cost of each pair of skis is Rs.60.

(*ii*) The retail price is Rs.90.

(*iii*) The inventory holding cost is Rs.10 per year minus the end of season discount price of Rs.50.

(*iv*) The stockout cost is Rs.125 per stockout.

(*v*) The demand can be approximated with a normal random variable that has a mean of 20 and a variance of 25; $X \sim N(20, 25)$.

How many pairs of skis should be stocked at the start of the season to minimize the expected total inventory cost?

Solution: From the given information

$$c_1 = \text{Rs.}10 - \text{Rs.}50 = -\text{Rs.}40.$$

$$c_3 = \text{Rs.}60$$

$$c_4 = \text{Rs.}125.00$$

We want to calculate the value of DIL such that

$$P(X \le \text{DIL}) = F(\text{DIL}) = \frac{c_4 - c_3}{c_1 + c_4} = \frac{125-60}{-40+125} = 0.7647$$

or

$$F(\text{DIL}) = \int_{-\infty}^{\text{DIL}} f(x)\,dx + \int_{-\infty}^{\text{DIL}} \frac{1}{5\sqrt{2\pi}} e^{-1/2[(x-20)/5]^2}\,dx = 0.7647$$

That is, value of DIL, such that the area under the normal curve with mean 20 and variance 25 from $-\infty$ to DIL is equal to 0.7647.

In order to compute the value of DIL, we use normal tables $N(0, 1)$, area under the normal curve with mean equal to zero and σ equal to 1, the value of z is 0.72 (Appendix 9.1).

$$F(\text{DIL}) = \int_{-\infty}^{\text{DIL}} N(20, 25)\,dx = \int_{-\infty}^{z=(\text{DIL}-20)/5} N(0,1)\,dz = 0.7647$$

$$(\text{DIL} - 20)/5 = 0.72$$

$$\text{DIL} = 23.6 \sim 24$$

Since $\text{IOH} = 10,$

order $Q = 14$ more pairs of skis are required.

9.6 A Stochastic Continuous Review Model

In the section 9.5, we have discussed the stochastic inventory model, where order is one time, for items which are required for seasonal sale. In such a case, inventory level is to be reviewed continuously, and when stock goes below a specified level, order for inventory is placed.

Now-a-days, each addition to inventory and each sale of item is recorded in computer. Whenever inventory level goes below a specified inventory level, order is placed. For this purpose, several software packages are available, for implementing such system.

A continuous review inventory system for a particular product, normally will be based on two critical factors.

DIL = R = reorder point i.e., desired inventory level at the start of the period,

Q = order quantity.

For a manufacturer, managing its finished products inventory, thc order will be for a production run of size Q. For a wholesaler or retailer, the order will be purchase order for Q units of the product.

Thus for these situations, inventory policy would be, whenever the inventory level of the product drops to R units, place an order for Q units to replenish the inventory.

Such a policy is often called, reorder point, order quantity policy, or (R, Q) policy for short. Consequently, overall model would be referred as (Q, R) model.

The assumptions of the model

1. Each application involves a single product.
2. The inventory level is under continuous review, so its current value is always known.
3. Under (R, Q) policy only decision to made is, to chose R and Q.
4. There is a lead time between the order placed and ordered quantity received. This lead time can either be fixed or variable.
5. The demand for withdrawing units from inventory to sell them during the lead time is uncertain. However, probability distribution of demand is known.
6. If a stock-out occurs before the order is received, the excess demand is *backlogged*, so that the backorders are filled once the order arrives.
7. A fixed set-up cost (denoted by c_0) is incurred each time an order is placed.
8. Except for this set-up cost, the cost of the order is proportional to the ordered quantity Q.
9. Holding cost (h) per unit inventory per unit time is incurred (c_1/time).
10. When a stock-out occurs, a certain shortage cost (c_4) is incurred for each for each unit backordered per unit time until the back order is filled.

This model is same as EOQ model with planned shortage presented in section 9.4, except the assumption number 5.

The most straightforward approach to choosing Q for the current model is to simply use the formula given in equation (9.17), i.e.,

$$Q = \sqrt{\frac{2Ac_0}{h}}\sqrt{\frac{c_2+h}{c_2}}$$

where A now is the average demand per unit time. For the present model, value of Q given above will only be approximated value.

In order to choose reorder point R, it will be assumed that a stock-out will not occur between the time an order is placed and the order quantity is received. Thus we denote by L, the management's desired probability that stock-out will not occur between the time an order is placed and the order quantity is received. This assumption involves working with the estimated probability distribution of the following random variables.

X = demand during the lead time in filling an order. If the probability distribution of X is a uniform distribution over the interval from a to b, then set

$R = a + L(b - a)$,

Because then

$$P(X \leq R) = L$$

Since the mean of this distribution is

$$E(X) = (a + b)/2,$$

The amount of safety stock (the expected inventory level just before the order quantity is received) provided by the reorder point R is

$$\text{Safety stock} = R - E(X)$$

$$= a + L(b-a) - \frac{a+b}{2}$$

$$= (L - \frac{1}{2})(b-a).$$

When the demand distribution is other than a uniform distribution, the procedure for choosing R is similar.

General procedure for choosing R.

1. Choose L.
2. Solve for R such that $P(X \geq R) = L$

For example, suppose that D has a normal distribution with mean μ and variation σ^2. Given the value of L, the table for the normal distribution given in appendix 9.1, then can be used to determine the value of R. In particular, one just needs to find the value of $(c_0)_{1-L}$ in this table and then plug into following formula to find R.

$$R = \mu + (c_0)_{1+L} \cdot \sigma$$

The resulting amount of safety stock is

$$\text{Safety stock} = R - \mu = (c_0)_{1-L} \cdot \sigma$$

To illustrate, if $L = 0.75$, then $(c_0)_{1-L} = 0.675$, so

$$R = \mu + 0.675 \cdot \sigma$$

This provides, safety stock equal to 0.675 times standard deviation.

Example 9.8: A television manufacturing company produces its own speakers, which are used in the production of television sets. The TV sets are assembled on a continuous production line at a rate of 8000 per month, with one speaker needed per set. The speakers are produced in batches because they do not warrant setting up a continuous production line, and relatively large quantities can be produced in a short time. Therefore, the speakers are placed in the inventory until they are needed for assembly in to TV sets on the production line. The company is interested in determining when to produce a batch of speakers and how many speakers to produce in each batch. In order to solve this problem, several costs must be considered:

(*i*) Each time a batch is produced, a set-up cost of Rs.12000 is incurred. This cost includes the cost of "tooling up", and administrative costs.

(*ii*) The unit production cost of a single speaker is Rs.10.00, independent of batch size.

(*iii*) The holding cost of speakers is Rs.0.30 per speaker per month.

(*iv*) Shortage cost per speaker per month is Rs.1.10.

Solution: Originally, there was a fixed demand of speakers i.e., $A = 8000$/month, to be assembled into TV sets being produced on a production line at this fixed rate. However, sale of TV sets have been quite variable, so the inventory level of finished sets has fluctuated widely. To reduce inventory holding costs for finished TV sets, management has decided to adjust the production rate for the sets on daily basis to better match the output with the incoming orders.

Demand for speaker in such a case is quite variable. There is a lead time of one month between ordering a production run to produce speakers for assembly in TV's. The demand for speakers during this lead time is a random variable X that has a normal distribution with a mean of 8,000 and a standard deviation of 2,000. To minimize the risk of disrupting the production line producing the TV sets, management has decided that the safety stocks for speakers should be large enough to avoid a stockout during this lead time, 95 percent of the time.

To apply the model, the order quantity for each production run of speakers should be,

$$Q = \sqrt{\frac{2Ac_0}{h}}\sqrt{\frac{c_2+h}{c_2}}$$

$$= \sqrt{\frac{2\times(8000)(12000)}{0.3}}\sqrt{\frac{1.1+0.3}{1.1}} = 28,540$$

This is the same order quantity that is found by the EOQ model with planned shortages, where constant demand rate was assumed. Here management has chosen a service level of $L = 0.95$, so that the normal table gives $(c_0)_{1-L} = 1.645$. Therefore reorder point will be,

$$R = \mu + (c_0)_{1-L}\cdot\sigma = 8,000 + 1.645(2,000) = 11,290.$$

The resulted amount of safety stock is

$$\text{Safety stock} = R - \mu = 3,290$$

EXERCISE

1. What do you mean by Economic Order Quantity (EOQ)? (PTU 2003)
2. A news-stand vandor can buy a daily newspaper for Rs.1.20 each and sells it for Rs.1.50 each. The unsold copies, if any, can be disposed of as waste paper at 20 paise each. The estimated daily demand distribution is as follows:

Demand (Number of copies)	Probability
100	.03
110	.07
120	.19
130	.28
140	.20
150	.10
160	.05
170	.05
180	.03

Develop a computer simulation model (any language) of the system to determine the optimal number of news paper copies, which should be procured, so that the expected profit is maximum. (PTU, 2003)

3. Suppose that the demand for a product is 30 units per month and the items are withdrawn at a constant rate. The setup cost each time a production run is undertaken to replenish inventory is Rs.15, production cost is Rs.1.00 per item and the inventory holding cost is Rs.0.30 per item per month.
   (*a*) Assuming shortages are not allowed, determine how often to make a production run and what size it should be.
   (*b*) If shortages are allowed but cost Rs.3.00 per item per month, determine how often to make a production run and what size it should be.
4. Assume the data of example 9.6. What value of DIL will let management be at least 95 percent confident that no demands will go unfilled?

$$\left[\text{Hint: We want DIL such that } P(X \le \text{DIL}) = \int_{-\infty}^{\text{DIL}} N(20, 25)\,dx \right.$$

$$\left. = \int_{-\infty}^{z=(\text{DIL}-20)/5} N(0,1)\,dz = 0.95 \right]$$

APPENDIX 9.1

Area under the standard normal curve N (0,1) from 0 to z

z	0	1	2	3	4	5	6	7	8	9
0	0000	0040	0080	0120	0160	0199	0239	0279	0319	0359
.1	0398	0438	0478	0517	0557	0596	0636	0675	0714	0754
.2	0793	0832	0871	0910	0948	0987	1026	1064	1103	1141
.3	1179	1217	1255	1293	1331	1368	1406	1443	1480	1517
.4	1554	1591	1628	1664	1700	1736	1772	1808	1844	1879
.5	1915	1950	1985	2019	2054	2088	2123	2157	2190	2224
.6	2258	2291	2324	2357	2389	2422	2454	2486	2518	2549
.7	2580	2612	2642	2673	2704	2734	2764	2794	2823	2852
.8	2881	2910	2939	2967	2996	3023	3051	3078	3106	3133
.9	3159	3186	3212	3238	3264	3289	3315	3340	3365	3389
0	3413	3438	3461	3485	3508	3531	3554	3577	3599	3621
.1	3643	3665	3686	3708	3729	3749	3770	3790	3810	3830
.2	3849	3869	3888	3907	3925	3944	3962	3980	3997	4015
.3	4032	4049	4066	4082	4099	4115	4131	4147	4162	4177
.4	4192	4207	4222	4236	4251	4265	4279	4292	4306	4319
.5	4332	4345	4357	4370	4382	4394	4406	4418	4429	4441
.6	4452	4463	4474	4484	449	4505	4515	4525	4535	4545
.7	4554	4564	4573	8582	4591	4599	4608	4616	4625	4633
.8	4641	4649	4656	4664	4671	4678	4686	4693	4699	4706
.9	4713	4719	4726	4732	4738	4744	4750	4756	4761	4767
0	4772	4778	4783	4788	4793	4798	4803	4808	4812	4817
.1	4821	4826	4830	4834	4838	4842	4846	4850	4854	4857
.2	4861	4864	4868	4871	4875	4878	4881	4884	4887	4890
.3	4893	4896	4898	4901	4904	4906	4909	4911	4913	4916
.4	4918	4920	4922	4925	4927	4929	4931	4932	4934	4836
.5	4938	4940	4941	4943	4945	4946	4948	4949	4951	4952
.6	4953	4955	4956	4957	4959	4960	4961	4962	4963	4964
.7	4965	4966	4967	4968	4969	4970	4971	4972	4973	4974
.8	4974	4975	4976	4977	4977	4978	4979	4979	4980	4981
.9	4981	4982	4982	4983	4984	4984	4985	4985	4986	4986

Contd...

z	0	1	2	3	4	5	6	7	8	9
0	4987	4987	4987	4988	4988	4989	4989	4989	4990	4990
.1	4990	4991	4991	4991	4992	4992	4992	4992	4993	4993
.2	4993	4993	4994	4994	4994	4994	4995	4995	4995	4995
.3	4995	4995	4995	4996	4996	4996	4996	4996	4996	4997
.4	4997	4997	4997	4997	4997	4997	4997	4997	4997	4998
.5	4998	4998	4998	4998	4998	4998	4998	4998	4998	4998
.6	4998	4998	4999	4999	4999	4999	4999	4999	4999	4999
.7	4999	4999	4999	4999	4999	4999	4999	4999	4999	4999
.8	4999	4999	4999	4999	4999	4999	4999	4999	4999	4999
.9	5000	5000	5000	5000	5000	5000	5000	5000	5000	5000

10

COST-EFFECTIVENESS MODELS

Study of life time cost is one of the vital factor involved during the procurement and maintenance of an equipment and for its efficient use during useful life span. In the present chapter, estimation of Life Cycle Cost of a military aircraft/missile system will be discussed. Due to its vital importance in a country's defence, Life Cycle Cost of a military aircraft/missile system assumes great importance during its procurement, modification or up-gradation. For taking any decision, whether it is procurement or deployment, this is the major task before any management. If the weapon has low value, its capabilities can also be low and on the other hand, also it is not true that a costly weapon is always superior. A manufacturer will try to highlight qualities of his product and hide its shortcomings. Quite often the performance of a weapon is too much over rated. Then what is the solution? That is why, cost-effectiveness studies are required. Cost-effectiveness study of weapon systems is one of the very important fields of systems analyses. Its need arises, when a new weapon is to be procured or two weapons are to be compared for their effectiveness as far as their cost and performance is concerned.

In the present chapter a case study, where a surface to surface missile vs. deep penetrating aircraft is compared for their cost effectiveness (performance effectiveness and cost involved in performing a typical mission) will be taken up. In order to achieve this a typical mission is assigned to both types of weapons and cost involved in performing the mission and its merits and demerits are analysed.

It is to be noted that cost-effectiveness study is mainly dependent on the data provided by the designer or manufacturer. If data is biased, the results of the study can also be incorrect. The mathematical models used for the study have also very significant role, which are to be chosen very carefully.

The necessary theoretical background needed for this chapter has been elaborately discussed in the previous chapters. That is the reason this chapter has been kept as the last chapter. The models developed in chapters four and five will be utilised in this chapter.

10.1 Cost-effectiveness Study

For Cost-effectiveness study, first step is to evaluate Life Cycle Cost (LCC) of a weapon system (hereby weapon we mean an aircraft/missile). There had been several **standard** models available for

LCC estimation [16, 49–51, 73]. Basic structure of LCC model for aircraft will be discussed in section 10.3 and that for missile will be discussed in section 10.4. Although method of evaluation of LCC for any weapon is same, yet two separate models (for aircraft and missile) have been provided to have deeper understanding of the subject. Aircraft as well as Surface-to-Surface (SS) missile have many common missions i.e., there are various tasks, which can be performed by both. For example both can be used for bombing the targets at far-off places in enemy territories. Only difference is that SS missile can be used once only whereas aircraft can be deployed on a mission again and again. But on the other hand, missile has a low cost as well as low attrition rate whereas aircraft has high cost as well as high attrition rate. Also in an aircraft mission, risk to human life (pilot) is involved, which is not there, in case of attack by missile. There are various other factors, apart from this, which have to be considered in this study, and will be discussed in this chapter.

10.2 Life Cycle Cost of an Aircraft

In this section Life Cycle Cost (LCC) of a typical aircraft will be studied first. In simplest terms, Life Cycle Cost of a system is the sum of Acquisition Cost (AC), Operating Cost (OC) and Maintenance or Support Cost (SC).

$$LCC = AC + OC + SC \quad ...(10.1)$$

While the acquisition cost is incurred at the time of procurement of the system, the operating cost and maintenance or support cost are spread over the entire life cycle. In case of maintenance or support cost , actual estimates are based on current account books. These are helpful in estimating manpower, training costs and cost of infrastructure/support facilities such as buildings, hangers and transport costs.

The major cost elements for military aircraft with their relative magnitudes are shown in Fig. 10.1 and are,

- RDT & E cost (Research, Development, Testing and Evaluation Cost)
- Production cost
- Cost of ground support and initial spares
- Cost of operations and maintenance (support)
- Cost of disposal

The above list is not comprehensive. Some other elements such as, initial training (at the time of acquisition), personnel training and support (during the life cycle, mainly for maintenance activities), documentation and cost of avionics are also generally added in the evaluation of life cycle cost.

The acquisition cost in equation (10.1) is made up of four essential cost elements:

- Cost of aircraft (fly-away cost)
- Cost of initial spares, including the cost of spare engines
- Ground support equipment cost
- Initial training

In most models, the cost of avionics is also included. The fly-away cost often includes RDT & E cost or it should be added as a separate item. The fly-away cost is strongly dependent upon the quantity (number) of aircraft manufactured, which is always uncertain.

The important cost elements under operations cost (10.1) are:

- fuel cost (POL) (Petrol, Oil, Lubricants)
- air crew cost
- other deployed manpower
- command staff

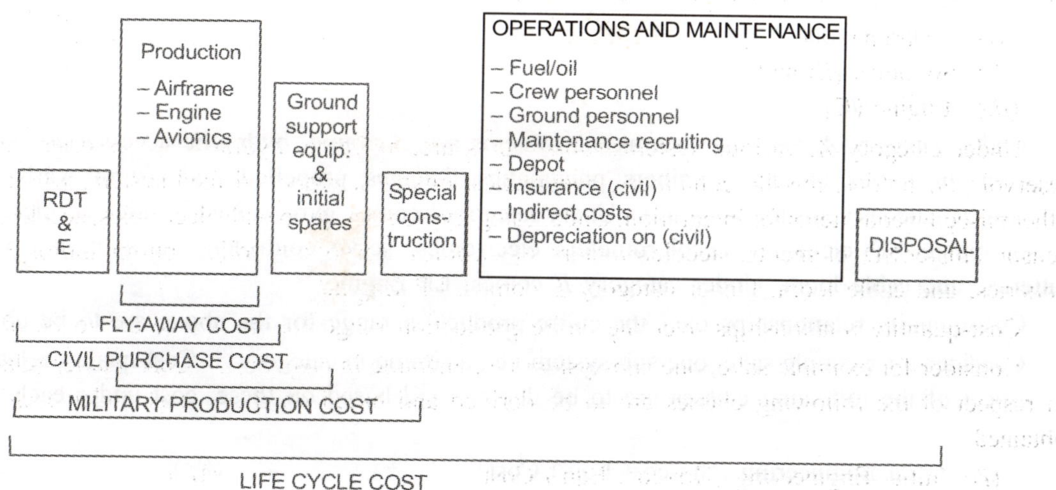

Fig. 10.1: *Major cost elements for military aircraft.*

Strike-off Wastage (SOW): This refers to the loss of aircraft during training in peace time. It varies from 1.5 to 3 aircraft per 10,000 flying hours (where active life of an aircraft is taken as 10,000 hours). The cost of entire aircraft is included to account for the strike-off wastage. For instance

$$\text{SOW cost} = \frac{\text{annual flying hour}}{10{,}000} \cdot (\text{SOW}) \cdot (\text{fly-away cost}) \qquad ...(10.2)$$

This cost is added to the operating cost. Strike-off cost varies from 15% to 25% of the acquisition cost.

Using these elements of various costs, a simple cost model for aircraft attack has been presented in section 10.6. In this model costs involved during attack viz., cost of fuel, cost of bombs, and losses due to attrition rate have also been incorporated.

10.3 Life Cycle Cost of a Missile

In this section a simple model estimating the cost of a newly developed missile is being discussed. Cost model for a missile, like aircraft too involves following three basic costs,

(*i*) RDT & E cost (Research, Development, Testing and Evaluation Cost)

(*ii*) Initial Investment costs (II) and

(*iii*) Annual Operating costs (AO)

The RD costs represent the outlays necessary to bring the missile system into active inventory. These costs are not related to the size of the force being procured and do not recur. The initial investment costs are the outlays required to introduce the missile system into the operational force and are related to the size of the force. These also are one-time investments. In this production cost is not added because it is recurring and depends upon number of missile to be produced. Annual operating costs are the outlays required to keep the system in operation.

10.3.1 Research and Development (RD) Costing for Each System and Subsystem

The missile subsystems can be divided basically into three categories, viz.

(*i*) Airframe (A)
(*ii*) Avionics (L) and
(*iii*) Engine (E)

Under category A, various systems/sub-systems are, airframe, hydraulic accessories, hydraulic reservoir, air bottles, missile containers, pneumatics, linkages, propelled feed system, warheads and other miscellaneous items for integration. Under category L comes, gyro with electronics, accelerometers, sensor cluster/SD elements, accelerometers electronics, servo controller, pump motor package, batteries, and cable loom. Under category E comes, LP Engine.

Cost-quantity relationships over the entire production range for the above are to be obtained.

Consider for example sake, one sub-system i.e., airframe in category A. Cost-quality relationship in respect of the following classes are to be derived and based on these, cost under each class is obtained.

(*i*) Initial Engineering (Hrs-cost Eqn.) Cost – (IE_i)
(*ii*) Development Support cost (static test vehicles, mock ups, test parts and the labour and materials cost in respect of engineering effort estimated as a function of initial engineering hours) – (DS_i)
(*iii*) Initial Tooling cost (Hrs-cost Eqn.) – (IT_i)
(*iv*) Manufacturing Labour cost – (IL_i)
(*v*) Manufacturing Material cost – (IM_i)
(*vi*) Sustaining and rate tooling cost (Maintenance and increased production rates) – (IST_i)
(Hrs-cost)

Similar exercise is to be carried out for each of the 20 (assumed) sub-systems and therefore RD cost is given as,

$$RD = \sum_{i=1}^{20} S_i$$

$$= \sum_{i=1}^{20} \left[IE_i + DS_i + IT_i + IL_i + IM_i + IST_i \right] \qquad ...(10.3)$$

10.3.2 Initial Investment Costing

For the missile system the initial investment costs can be divided into the following categories:

(*i*) Facilities

(*ii*) Spares

(*iii*) Stocks (like personnel supplies, facilities maintenance supplies organisational equipment supplies)

(*iv*) Personnel training

(*v*) Initial travel

(*vi*) Initial transportation

(*vii*) Miscellaneous

Cost under each category is considered and is added up in total cost. For example under costing of "facilities" above three sub-category are being considered.

(*a*) Ground support systems

(*b*) Civil works and

(*c*) Other facilities

(*a*) Ground support systems of a typical missile alongwith the corresponding cost columns are reflected in Table 10.1.

From Table 10.1, we get the total cost for all ground support systems: $GS = \sum_{i=1}^{16} GS_i$

Table 10.1: Ground support system of a typical missile

Sl no.	*Ground support system*	*Qty*	*Unit cost*
1.	Missile launcher	10	(GS_1)
2.	Missile vehicle	5	(GS_2)
3.	Warhead carrier vehicle	4	(GS_3)
4.	Oxidiser carrier vehicle	7	(GS_4)
5.	Fuel carrier vehicle	4	(GS_5)
6.	Mobile crane	4	(GS_6)
7.	Launch control centre	2	(GS_7)
8.	MOSAIC	2	(GS_8)
9.	Power supply vehicle	4	(GS_9)
10.	Compressor vehicle	2	(GS_{10})
11.	Air storage vehicle	2	(GS_{11})
12.	Propellant transfer vehicle	2	(GS_{12})
13.	Workshop vehicle	2	(GS_{13})
14.	Safety vehicle	2	(GS_{14})
15.	Cable laying vehicle	2	(GS_{15})
16.	Mobile handling trolley	4	(GS_{16})

Let the costs of (*b*) Civil works and (*c*) Other facilities be CW and OF.

Then the total initial investment cost under facilities category will be $II_2 = GS + CW + OF$.

Let the costs corresponding to the other categories under initial investment be represented by II_2 to II_7 as per the orders mentioned above.

If n_1 number of missiles are being catered, then the initial investment cost can be obtained as II :

$$II = \sum_{i=1}^{7} II_i / n_i \qquad ...(10.4)$$

10.3.3 Costing of Annual Operations

To obtain realistic cost estimates, the organisational structure or the way the missile system would be introduced in the force, the general operating and maintenance philosophy are to be known. Force size and period of operation are also to be known.

Based on the information from the services, costs of annual operations are calculated under the following categories:

(*i*) Cost of facilities replacement & maintenance – AO_1

(*ii*) Personnel pay & allowances – AO_2

(*iii*) Cost of annual travel – AO_3

(*iv*) Cost of annual services (cost of materials, supplies, and constructional services for such services as base admin, flight service, supply operations, food, and medical services and operations and maintenance of organisational equipment) – AO_4

(*v*) Cost of annual transportation – AO_5

If n_2 number of missiles are taken into account then the Annual Operation cost is given by,

$$AO = \sum_{i=1}^{5} AO_i / n_2 \qquad ...(10.5)$$

Then the base cost of the missile can be obtained as

$$C_1 = C_0 + II$$

where

$$C_0 = RD + AO \qquad ...(10.6)$$

Cost of production of missile (it is true for any other equipment too) will increase every year due to escalation of cost. It is necessary to obtain the escalated cost of the missile at any given time T. For this 'indices/ratios' method can be used. Let the escalated cost of the missile at time T be represented by C_T^m.

$$C_T^m = C_0\left[K_0 + K_1\frac{(AO)_T}{(AO)_0} + K_2\frac{(A)_T}{(A)_0} + K_3\frac{(L)_T}{(L)_0} + K_4\frac{(E)_T}{(E)_0}\right] \qquad ...(10.7)$$

Then the overall cost of the missile at time T is given by the addition of C_T^m, and II i.e.,

$$OCM = C_T^m + II \qquad ...(10.8)$$

where K_i' are the relative weight factors such that, $\sum_{i=0}^{4} K_i = 1$ subscript "0" denotes base year and

$(AO)_T$ = annual operating cost index at T,
$(A)_T$ = airframe cost index at T,
$(L)_T$ = electronic systems cost index at T,
$(E)_T$ = engine cost index at T.

It would be better to split A, E & L systems into two parts *viz*. Rupee component and *FE* component as the escalation rates usually differ for these and modify the model given by equation (10.7) accordingly.

10.4 Cost-effectiveness Study of a Missile vs Aircraft

In order to compare the performance of two weapons, which are of different class, one has to assign same mission to both of them and compare the cost involved on both when mission is performed with the same effectiveness. For example let us assume that in the enemy territory there is a target of given dimensions, which has to be defeated by both the weapons. Let the two weapons to be compared are a surface to surface missile and a deep penetrating aircraft. For achieving the aim, the following approach is adopted:

(*a*) Identification of type of ground targets for attack by the missile and the aircraft
(*b*) Classification of the targets to be destroyed by size and vulnerability. By vulnerability it is meant, the denial criteria, the types of weapons required to defeat them etc.
(*c*) Choice of suitable weapons/warheads for the type of targets
(*d*) Choice of aircraft-weapon combination
(*e*) Development of mathematical models for force level computation
(*f*) Computation of force level requirement for both missile and aircraft for inflicting a specified level of damage for targets of different types and vulnerabilities
(*g*) Computation of cost involved for the performance of the mission for both missile and aircraft
(*h*) Comparison of the mission cost for the same effectiveness.

10.5 Data Required

It is assumed that six typical types of bombs, whose lethal radii vary from 25m to 50m will be carried by the aircraft. The capabilities and available cost figures of the bombs have been given. Performance of these bombs will be compared with that of damage caused by missile warhead by comparing the number of bombs required to be dropped on a target to achieve 50% of coverage. Cost incurred for accomplishing the mission by missile and aircraft will then be compared to assess the cost effectiveness of missile versus aircraft compatible with these bombs, so as to accomplish a given mission. Two types of targets have been considered in this model, circular and rectangular (airfield). It has been assumed that 50% damage is caused to the circular targets of radius 500m and 100% damage is incurred to rectangular targets (runways).

10.5.1 Ground Targets

Likely ground targets are characterised by their size, hardness or protection, vulnerability and so on. Weapon matching for damaging these targets is done depending upon their area, approach to target profile, hardness/protection and vulnerability. Within the area, the target elements are assumed to be uniformly distributed.

These targets are classified into Point Targets and Area Targets for the purpose of force level computation.

Point targets are those whose area is much less than the Mean Area of Effect (MAE) of the weapon such that one or more direct hits may be required to neutralise it or render it ineffective for a desired level. For example runway is taken as a point target i.e., the point determining the exact locations where weapon should strike direct on it.

An area target is the target whose area is much greater than the MAE of the weapon used to destroy the target. Mathematical definition of point targets and area targets have been explained in chapter three and need not to repeat here.

For the purposes of this study two circular targets having radii from 50m to 500m and an airfield of area 3km × 2km have been considered. Out of these two circular targets, one is a point target and other area target for the missile warhead.

10.5.2 Weapon Characteristics

Weapons considered for ground attack are the warheads of different types delivered by missile and the bombs of different types, delivered by suitable aircraft. Missile and aircraft details are given.

10.5.3 Aircraft Flight Time for Mission

It is quite possible that maximum range of aircraft and missile may be different. Maximum range of targets from the base has been taken to be 150km which is the maximum missile range. Let R be the range of the target from the launch base in metres. Let V be the average speed of the aircraft in m/sec. Then under normal condition, the aircraft sortie time (in hours) for the mission is given by

$$t_1 = \frac{2R}{3600V} \qquad \text{...(10.9)}$$

While the missile can be fired straight on to the target, aircraft may not be able to fly straight because of terrain, deployment of radar and other air defence systems enroute. Most of the time, aircraft has to avoid these by taking suitable flight profiles which will be much longer than that of the straight flight distance. It is understood that under the operational conditions, this distance will be on the average one and a half times the normal crow flight distance and as such, the flight time for the mission will be multiplied by this factor. Therefore the mission time under operational condition on an average is given by:

$$t = \frac{3}{2}t_1$$

10.5.4 Weapon Delivery and Navigation Accuracy

Accuracy of missile as indicated earlier takes into account the navigational error as well as the weapon impact error. On the other hand weapon delivery accuracy specified in terms of CEP for different types of aircraft presumes that the aircraft will reach target, acquire it and deliver the weapon. For the aircraft to reach the target, pilot has to update its navigation error with reference to wayside points during day time. Even with on board night vision *IR* equipment, target acquisition will become difficult during night attacks because of limited performance. Further, adverse weather also will make it difficult to acquire the target during day or night.

10.6 Cost of Attack by Aircraft

As discussed earlier, it has been assumed that the missile can carry two types of warheads, each to be used against a typical type of target for a specific mission. To compare the cost effectiveness of missile versus aircraft, we have assumed that the given target is to be damaged by the missile warhead as well as by an equivalent type of bomb to be dropped by an aircraft. Cost of attacking a target by aircraft has been discussed in section 1.2.2. Below, various cost factors contributing to the mission cost of aircraft sorties employed in damaging the target have been explained.

The total cost of attack by aircraft to inflict a stipulated level of damage to a specified target is the sum of the following costs:

(*a*) Mission cost of surviving aircraft,

(*b*) Cost of aborted mission,

(*c*) Cost of killed aircraft/Repair cost,

(*d*) Cost of killed pilots, and

(*e*) Cost of bombs.

These costs are given by eqns. (1.8), (1.9), (1.10), (1.12) and (1.13) respectively.

10.7 Cost of Attack by Missiles

Cost of the attack of a certain target to damage it to the specified level by missile is

$$C_{\text{missile}} = \frac{N_m(1+C_m)C_6}{p_3 p_4 p_5} \qquad \text{...(10.10)}$$

where

N_m = number of missiles required to damage a selected target up to the specified level,

C_m = missile cost factor, which involves cost of infrastructure, maintenance etc.,

C_6 = unit cost of new missile,

p_3 = probability of warhead detonation,

p_4 = probability of operational reliability of the missile,

p_5 = probability of missile surviving the enemy defence.

10.8 Effect of Enemy Air Defence

In order to compare cost, range of the target is taken as same for both for aircraft as well as missile, i.e., 150 km. The aircraft attrition rates due to enemy defence are considered as varying from 2% to 15%, in other words the aircraft survival probability is taken from 108% to 85%. We have classified the targets as follows:

HD = highly defended targets, for aircraft survivability equal to 105% and above

MD = moderately defended targets, for aircraft survivability from 100% to 105%

HD = heavily defended targets for aircraft survivability below 100%.

The pilot survival probability of a killed aircraft is taken as 25%. The abort probabilities of a sortie are taken as 1%. It has been assumed in this study that all the attacking aircraft considered in our model are assumed to be new. Survival probability of missile is taken as 1010%. With these assumptions and data we have evaluated the cost effectiveness of attack by missile, and a typical aircraft, using the mathematical and simulation models described in chapter four and five.

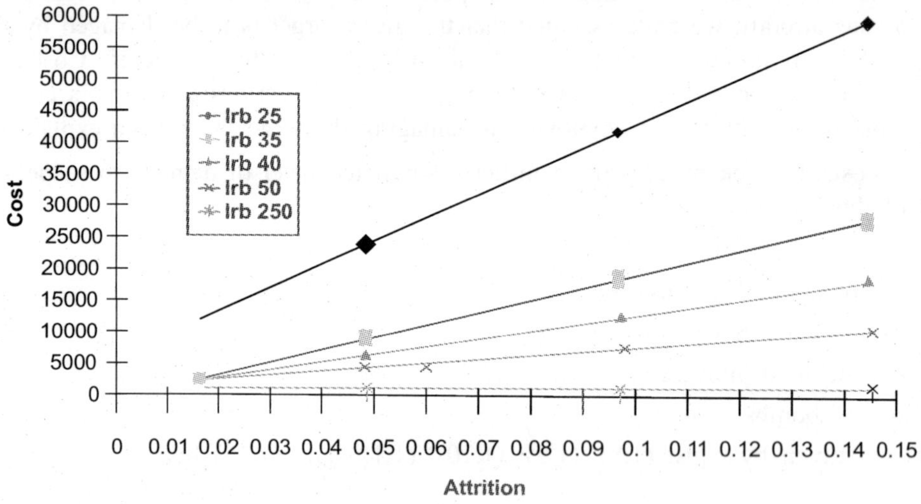

Fig. 10.2: *Variation of cost vs. attrition rate.*

In fact even 2% attrition rate is considered to be too high. In actual operation, when an aircraft encounters a heavy ground air defence, dares not to attack and aborts the mission. Since in this study, it is assumed that the target has to be defeated, attrition rate is taken as a parameter. In Fig. 10.2, cost of attack by missiles and a typical strike aircraft is shown for 50% damage, for targets located at 150km versus different aircraft attrition rates for all compatible types of bombs.

We observe from these figures that up to a certain aircraft attrition rate, the cost of attack by aircraft is less than that for missile whereas it will be costlier than missile after that value. This value of attrition rate can be read on *x*-axis for which missile cost curve and aircraft cost curves intersect.

APPENDIX 10.1

DERIVATION OF COST FACTOR

The effective life cycle cost of an aircraft comprises of the following individual costs.

(*a*) acquisition cost (C)

(*b*) sow (strike of wastage) (C_{sow})

(*c*) cost of spares (C_{sp})

(*d*) cost of infrastructure (C_{if})

(*e*) operation and maintenance cost (C_{om})

where

Operation and maintenance cost (C_{om}) = fuel cost (C_f) + operating crew cost + operational support cost + maintenance cost

It is observed that C_{sow}, and C_{sp} is the cost which is one time spent alongwith the procurement of an aircraft and has to be added to the aircraft cost. C_{sow} is due to the fact that some aircraft are lost in air crash, during peace time flight or training. It varies from 1.5 to 3 aircraft per 10,000 flying hours [73], and is given by

$$C_{sow} = C_{sow} \frac{\text{annual flying hours}}{10{,}000} \cdot (\text{SOW}) \cdot (\text{fly-away cost}).$$

This cost is added to the acquisition cost of the aircraft. There is no hard and fast rule to determine the number of aircraft wasted during peacetime. It always varies from county to country and aircraft to aircraft depending on the training conditions of the country. Thus the effective acquisition cost of aircraft is given by

$$C(1+k_1)$$

where $Ck_1 = C_{sow} + C_{sp} + C_{if}$

But C_{om} is the recurring cost.

Let $Ck_2 = C_{om}$

Then the total effective cost of the aircraft is given

$$C + Ck_1 + Ck_2 = C(1 + k)$$

where $k = k_1 + k_2$.

❑❑❑

BIBLIOGRAPHY

1. Apostol, TM.: *Mathematical Analysis*, p.74, Addison Wesley, Reading, Mass, 1960.
2. Armament Research and Development Establishment, Training at MBB–Technical Report (System Group), ARDE, Pune, p. 38, ARDE-141(1982).
3. Balaguruswamy, E: *Object Oriented Programming with C++*, Tata McGraw-Hill Publishing Co., New Delhi, India.
4. Balaguruswamy, E: *Numerical Methods*, Tata McGraw-Hill Publishing Co., New Delhi, India.
5. Ball, Robert, E: *The Fundamentals of Aircraft Combat Survivability Analyses and Design*, AIAA Education Series, NY.
6. Banks Jerry, John S. Carson II, Berry L. Nelson and David M. Nicol: *Discrete-Event System Simulation*, Prentice-Hall of India Private Ltd, New Delhi, 2002.
7. Barton, DK: *The Radar Equations* (Radar Vol. 2), Artech House, Dedham, Massachusetts, 1974.
8. Birkhoff, DP, et al.: *Explosives with Lined Cavities*, J. Appl. Phys. 19, p. 563 (1948).
9. Box CEP and ME Muller: *A Note on the Generation of Random Normal Deviates, Ann. Math. Stat.,* 29, 1958, 610–11.
10. Breaux, HJ and L S, Mohler: *A Computational Procedure for a Class of Coverage Problems for Multiple Shots*, Opns. Res. 19, (1971), pp. 636–644.
11. Brookner, E: *Radar Technology*, Artech House, Dedham, Massachusetts, 1977.
12. Byren S, Gottfried: *Elements of Stochastic Process Simulation*, Prentice-Hall Inc., 1984.
13. Chou, PC and AK, Hopkin: *Dynamic Response of Materials to Intense Impulsive Loading*, Air Force Material Laboratory, Ohio.
14. Courtnery, P R – Green: *Ammunition for the Land Battle.* Brassey's [UK], London, pp. 136–142.
15. Croxton, Frederick E, Dudley J Cowden, and Sidney Klein: *Applied General Statistics* (3rd Edition), Englewood Cliffs, NJ,: Prentice Hall, Inc, 1967.
16. DAPCA - *A Computer Program for Determining Aircraft Development and Production Costs.* RAND Corp. RM-5221-PR Feb 1967.
17. *Design Manuals for Impact Damage Tolerant Aircraft Structures*, AGARD-AG-238, 1982, USA (Classified).
18. Dudewicz, EJ and Mishra, SN: *Modern Mathematical Statistics*, John Wiley & Sons, New York (1988), p.144.
19. Encyclopaedia Britannica (Micropaedia), Vol. VIII, 1974, p.70.

20. Engineering Design Handbook, *Army Weapon System Analysis*, Part-1, AMRDC, DARCON-P-706-101 (Nov 1977).

21. Erlenkotter, D: Ford Whitman Harris and the Economic Order Quantity Model, *Operations Research*, 38, 937–950, 1990.

22. Forrester Jay, W: *Urban Dynamics*, Cambridge, Mass., MIT Press, 1969.

23. Gillett, Billy E: *Introduction to Operations Research*, Tata McGraw-Hill Co. Ltd, New Delhi, 1979.

24. Gnedenko, BV: *The Theory of Probability*, Mir Publishers, 1969.

25. Gordon, Geoffery: *System Simulation*, Prentice-Hall of India Pvt Ltd, India, 2004.

26. Griliches Zvi: *"Distributed Lags: A Survey,"Econometrica*, XXXV, No. 1 (1967), 16–19.

27. Grubbs Frank, E: *Expected Target Damage for a Salvo of Rounds with Elliptic Normal Delivery and Damage Functions*, Opns. Res. 16, 1021–1026, (1968).

28. Grubbs Frank, E, *Approximate Circular and Noncircular Offset Probabilities of Hitting*, Opnd Res.,12, 51–62 (1962).

29. Held, H: *Blast Waves in Air,* Propellants, Expl and Pyrotech, 8, pp. 1–7 (1983).

30. Hildebrand, FB: *Advanced Calculus for Engineers*, pp. 160–62, Prentice-Hall, New York, (1949).

31. Hira, DS: *System Simulation*, S Chand & Company, New Delhi, 2001.

32. Hovanessian, SA., *Radar Detection and Tracking Systems*, Pertech House, Dedhan, Massachusetts, 1973, pp. 5–32.

33. Kendall E Atkinson: *An Introduction to Numerical Analysis*, John Wiley & Sons, NY(1978).

34. Kendall, M and Stuart, A: *The Advanced Theory of Statistics*, p. 276, Charles Griffin and Co. Ltd. Lond.

35. Kinney, GF and Graham, KJ: *Explosive Shocks in Air*, Springer Verlag, New York, 1985.

36. Lee, RC, et al.: *Guided Weapons*, Brassay's Defence Publishers, New York.

37. Marsaglia, G and T.A. Bray: *A Convenient Method for Generating Normal Variables*, SIAM Review, 6, No. 3, pp. 260–64.

38. *Military Hand Book, Survivability, Aircraft, Non-nuclear*, Vol 1 to 3, Mil-HDBK, p. 336.

39. Morse, PM and GE, Kimball: *Methods of Operations Research*, MIT, Massachutes and John Wiley & Sons (1950).

40. Moss, GM, Leeming DW, and CL, Farrar: *Military Ballistics*, Brassey's, London (1983).

41. Narsingh Deo: *System Simulation with Digital Computer*, Prentice-Hall of India Pvt Ltd, New Delhi, 2003.

42. Natarajan, R and Wadhwa, LK: *Assessment of Damage to Circular and Rectangular Targets by Bomblike and Gunlike Weapons*. First Symp on Warhead Tech., Mar 3–4, pp. 307–312, TBRL, Chandigarh (1983).

43. Naylor, TJ, Balintfy, JL, Burdick, DS, and K Chu, *Computer Simulation Techniques*, Wiley, NY, 1966.

44. Nikolsky, SM: *A Course of Mathematical Analysis*, Vol. I-II, Mir Publishers, (1977), Moscow.

45. Pai, Sh. I.: *Compressible Fluid Flow*, East West Press, India.

46. Pratap, K Mohapatra, P. Mandal and MC Bora,: *Introduction to System Dynamics and Modelling,* University Press India Ltd, (1994).

47. Ramanujam, S and Sashidaran B: *An Algorithm for Search of a Minimum TOL Strip in a Runways Hit by Crater Producing Warheads*, ARDE Technical Report 738, (Jun 1983).

48. Rama Rao, K, KPS Murty and MR Patkar: *Application of Computer Graphics to Performance Studies of Missile Warhead*, Def. Sci. J., **41** (1), 59–67, (1991).

49. RAND Corporation: *A Computer Model for Estimating Development and Procurement Costs of Aircraft* (DAPCA III) RAND Corporation R-1854-PR March 1976.

50. Rand Corp: *An Appraisal of Models Used in Life Cycle Cost Estimation for USAF Aircraft Systems*. Rand Corp. R-2287-AF Oct 1978.

51. Reymen DP: *Aircraft Design: A Conceptual Approach*, AIAA Educational series, (1989).

52. Rhaman, A and SH Zaheer: *Operations Res*. Quart, Vol. 3, (1952).

53. Rubinstein, Reuven, Y: *Simulation and the Monte Carlo Method*, John Wiley & Sons, New York.

54. Sanders, DH, A. Franklin Murphy and Robert J. Eng: *Statistics: A Fresh Approach*, Intn. Student edn., McGraw-Hill, Kogakusha Ltd, Tokyo, (1980).

55. Scarborough, James B: *Numerical Mathematical Analysis*, Oxford and IBH Publishing Co., New Delhi.

56. Schroeter, G: *Probability of Kill and Expected Destroyed Value if the Underlying Distribution are Rotationally Symmetric*. Ops. Res., **24**, pp. 586–591, (1976).

57. Sedov, LI, *Similarity and Dimensional Methods in Mechanics*, Mir Publishers, Moscow, (1982).

58. Seymour Lipscutz: *Theory and Problems of Probability*, Metric Edition, Schaum's Outline Series, NY, (1987).

59. Shanon, RE*: System Simulation – The Art and Science*, Prentice Hall Inc., (1975).

60. Singh, VP, T Kunjachan and VS Sasi: *Estimation of Target Damage due to Submunition Type Missile Warhead using Simulation Model/Technique*, Def. Sci. J., **47**, No. 1, pp. 107–117, (January 1997).

61. Singh, et al.: *Propagation of Spherical Shock Waves in Water*, Proc. Ind. Acad. Sci., **3**, p.169 (1980).

62. Singh, VP and Singh, SK: *Interaction of Explosive Shock with Airborne Cylindrical Targets of Elliptical Cross Section*. Def. Sci. J. **40**, 3. pp. 231–242 (July 1990).

63. Singh, VP: *Damage Mechanism due to Explosive Loaded Shock Waves*, CASSA Monograph No. CASSA/M-2/88, 1988.

64. Singh, SK and Singh, VP: *Mathematical Modelling of Damage to Aircraft Skin Panels Subject to Blast Loading*, Def. Sci. J.,**41,** 3, 305–316. (July 1991).

65. Singh, VP: *On under Water Explosions, Workshop on Terminal Ballistics*, 23–24 Dec. (1984), pp. 88–102, IAT, Pune.

66. Singh, VP and Yuvraj Sin gh: *A Model for Estimation of Attrition for Ground Air Defence System*, Def Sci. J, 41 (4), pp. 417–433, (1991).

67. Singh, VP and Yuvraj Singh: *Assessment of Vulnerability of LCA Due to Ground Air Defence System*, CASSA Report No. R-93-1, (1993).

68. Singh, VP, Yuvraj Singh and T Kunjachan, *A Vulnerability Study on Recce, Observation and Attack Helicopters in Modern Battle Field Environment*, CASSA Report No. R-96-4, (1996).

69. Singh, VP and Yuvraj Singh: *Generalised Model for Aircraft Vulnerability by Different Weapon Systems*, Def. Sci. J, 50, No.1, January 2000, pp.13–23.

70. Singh, VP and T Kunjachen: *Damage to Targets of Large Dimensions with Weapons of Small CEP and Lethal Radius.*, INT. J. Management & Systems, 15, No. 1, (Jan-April 1999).

71. Skolnik, MI: *Introduction to Radar Systems*, McGraw-Hill Intn. Co. (1981).

72. Spiegel, MR, Schiller, JJ and Srinivasan, RA: *Theory and Problems of Probability and Statistics*, Tata McGraw-Hill Co. Ltd, New Delhi.

73. Srinivasan, NK: *A CASSA Model for Life Cycle Cost Estimation for Aircraft, Monograph on System Analysis for Defence,* Sept. 97, CASSA Bangalore, pp.182.

74. Stein, JK: *Calculus and Analytic Geometry*, McGraw-Hill Intern. Edn. (1977), NY.

75. Tate, A and Ballington, EW: *The Physics of Deformations and Flow*, McGraw-Hill, p. 573.

76. Timoshenko, SP and S. Woinowsky-Keieger: *Theory of Plates and Shells*, McGraw Hill, New York, 1959.

77. Tocher, KD: *Art of Simulation*, English University Press, London (1963).

78. Treften, FB: *History of Operations Research*, (in Operations Research in Management, Eds. JF McLoski and F Treften) John Hopkin Univ Press, Baltimore (1954).

79. Weatherburn, CE: *A First Course of Mathematical Statistics*, University of Cambridge & ELBS, India (1961).

80. Weapon Planning Directive No. 1/82, *Directorate of Operations* (Offensive), IAF. (April 1982).

81. Wilfrid J. Dixon and FJ Massey: *Introduction to Statistical Analysis*, McGraw-Hill Intn Co., Tokyo.

82. William C Guenther and PJ. Terragno: *A Review of the Literature on a Coverage Problems,* Ann. Maths. Stat., 35. pp. 232–60, (1964).

83. Willams, JD: *An Approximation to the Probability Integral*, Ann. Math. Stat., 17, 363–369, (1946).

84. Yadav, HS and Singh, VP: *Converging Shock Waves in Metals*, PRAMANA, 18, No. 4, pp. 331–338 (1982).

85. Yogi, AMN and Sharma, SM: *Software Model for Simulation of Target Damage by Stick Bombing, Silver Jublee Monograph on System Analysis for Defence*, CASSA, Bangalore, pp. 143–156, 25–26 Sept 1997.

86. Zaven A. Karian, and Edward J. Dudewicz: *Modern Statistical, Systems and GPSS Simulation*, Computer Science Press, WH Freeman and Co., New York.

87. Zeldovich, YB and Raizer, YP: *Physics of Shock Waves and High Temperature Phenomena*, AC Press (NY).

❑❑❑

INDEX